Sven Landsgesell

Magneto-Electric Properties in Neodymium Yttrium Manganites

Sven Landsgesell

Magneto-Electric Properties in Neodymium Yttrium Manganites

How Orthorhombic Neodymium Yttrium Manganites Become Multiferroic

Südwestdeutscher Verlag für Hochschulschriften

Imprint

Any brand names and product names mentioned in this book are subject to trademark, brand or patent protection and are trademarks or registered trademarks of their respective holders. The use of brand names, product names, common names, trade names, product descriptions etc. even without a particular marking in this work is in no way to be construed to mean that such names may be regarded as unrestricted in respect of trademark and brand protection legislation and could thus be used by anyone.

Publisher:
Südwestdeutscher Verlag für Hochschulschriften
is a trademark of
Dodo Books Indian Ocean Ltd., member of the OmniScriptum S.R.L Publishing group
str. A.Russo 15, of. 61, Chisinau-2068, Republic of Moldova Europe
Printed at: see last page
ISBN: 978-3-8381-1931-1

Zugl. / Approved by: Berlin, TU, Diss., 2010

Copyright © Sven Landsgesell
Copyright © 2010 Dodo Books Indian Ocean Ltd., member of the OmniScriptum S.R.L Publishing group

Contents

1	**Introduction**		**3**
	1.1 Ferroic Materials		3
		1.1.1 Primary Ferroics	3
		1.1.2 Ferromagentism	4
		1.1.3 Ferroelectricity	5
	1.2 Magnetoelectric Multiferroics		6
		1.2.1 The Perovskite Structure	7
		1.2.2 Cooperative Jahn-Teller Distortion	9
		1.2.3 Magnetic Exchange in RMnO$_3$	10
	1.3 Motivation and Scope of This Work		16
2	**Experimental Methods**		**17**
	2.1 X-ray Diffraction		17
		2.1.1 Laue diffraction	19
	2.2 The Rietveld method		19
	2.3 Neutron Scattering		20
		2.3.1 Magnetic Neutron Scattering	22
	2.4 Magnetization Measurements		23
	2.5 Dielectric Measurements		24
	2.6 Pyroelectric Measurements		25
3	**Synthesis**		**27**
	3.1 Synthesis of Polycrystalline Nd$_{1-x}$Y$_x$MnO$_3$		28
	3.2 Single Crystal Growth of Nd$_{1-x}$Y$_x$MnO$_3$		28
		3.2.1 The Floating Zone Method	29
		3.2.2 Optimizing Conditions for Crystal Growth of Nd$_{1-x}$Y$_x$MnO$_3$	33

	3.2.3	Characterization of the Quality of Single Crystals	34
	3.2.4	Thesis objective	36

4 Characterization of Crystal and Magnetic Structure of $Nd_{1-x}Y_xMnO_3$ — 39
    4.1 Isotropic Magnetic Properites ... 39
    4.2 Room Temperature Crystal Structure ... 42
    4.3 Magnetic Structure ... 45
    4.4 Phase Diagram of the Polycrystalline Samples ... 50

5 Results of Physical Property Measurements — 53
    5.1 Systematic Changes of Ferroelectric Properties with x ... 55
    5.2 Results of Measurements on $x = 0.55$ sample ... 56
    5.3 Similarities and Differences to other RMnO$_3$ ... 60

6 Magnetic Order in $Nd_{1-x}Y_xMnO_3$ — 67
    6.1 Magnetization Measurements on Single Crystals ... 67
    6.2 Variation of Magnetic Order with x ... 68
    6.3 Magnetic Order in low Doped Region ($x = 0.30, 0.35$) ... 71
    6.4 Magnetic Order in Phase Co-Existence Region ($x = 0.40, 0.45$) ... 73
        6.4.1 The $x = 0.40$ region ... 73
        6.4.2 The $x = 0.45$ region ... 75
    6.5 Magnetic Order in the Higher Doped Region ($x = 0.50, 0.55$) ... 79
    6.6 Discussion of the Magnetic Order ... 79

7 Cycloidal Order in $Nd_{1-x}Y_xMnO_3$ — 83
    7.1 The Irreducible Representations of the Pbnm Space Group ... 83
    7.2 Magnetic Structure $x = 0.45$ and 0.55 ... 85

8 Summary and Conclusions — 91
    8.1 The $Nd_{1-x}Y_xMnO_3$ Magnetic Phase Diagram ... 91
    8.2 Multiferroic Properties ... 92
    8.3 Phase co-existance ... 94
    8.4 Conclusion ... 95

A Appendix A: Instruments Used — 99
    A.1 SQUID ... 99
    A.2 PPMS ... 100
    A.3 VSM ... 101
    A.4 Dielectric Measurement Device ... 101
    A.5 Polarization Measurement Device ... 102

A.6	E9	102
A.7	E4	102
A.8	E5	103
A.9	V2	104
A.10	Bruker	104
A.11	Laue	104
A.12	D10	105
A.13	D20	105
A.14	Image Furnace	106

B Appendix B: Detailed Figures   109
   B.1   Results of Physical Properties Measurements 109

C Appendix C: Publications   121

Bibliography   123

Chapter 1
Introduction

Der Mensch versteht nichts - aber das auf einem sehr hohen Niveau.

Dieter Nuhr

1.1 Ferroic Materials

1.1.1 Primary Ferroics

There are three primary ferroics, ferromagnets, ferroelectrics and ferroelastics, that show a phase transition, at a temperature T_C, where they undergo changes in the microscopic and macroscopic scale. In the microscopic scale the symmetry of the crystal is reduced, with the loss of one point group symmetry operator as for example by a microscopic displacement of atoms in a ferroelectric. To reduce the free energy, the ferroic crystal form domains, so that the ferroic properties averages over the entire crystal macroscopically to sum to zero[1].

An appropriate external field (magnetic field for ferromagnets, electric field for ferroelectrics and mechanical stress for ferroelastics) allows the domains to reorient in the same direction along the field leading to a persistent ferroic property when the field is withdrawn. When the external field is reversed the direction of a ferroic property also reverses forming a characteristic hysteresis curve that can be measured as shown in Figure 1.1.

If the order parameter of a ferroic phase transition is the ferroic property itself the ferroic is called a proper ferroic in contrast to improper ferroics where the ferroic property is consequence of an other process[2]. When two different primary ferroic properties are to be found in the same material it is called a secondary ferroic, like for example materials that show a magnetic and a ferroelectric order. If those two components are coupled to each other they are called a multiferroic, meaning in this particular case magnetoelectric.

To introduce the magnetoelectric multiferroics the following sections will briefly describe the properties of ferromagnetism, ferroelectricity and the magnetoelectricity, while discussion on the third class of ferroics, ferroelastics is omitted.

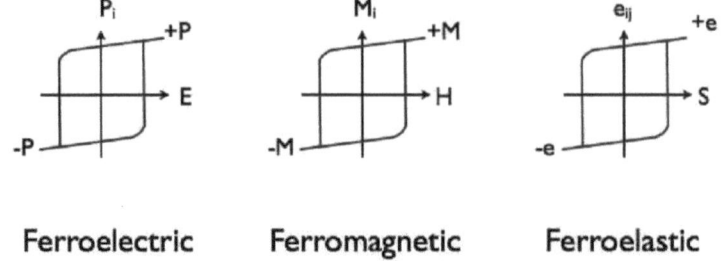

Figure 1.1: *A defining feature of ferroic materials is that they exhibit hysteresis. Here three typical hysteresis curves for a ferroelectric are shown (spontaneous polarization P under the application of an electric field E), a ferromagnet (spontaneous magnetization M under the application of a magnetic field H), and a ferroelastic (spontaneous elastic strain e_{ij} under the application of stress field S) material, respectively. The reversal of the sign of the applied field in the three cases causes a reversal in the direction of the ferroic property.*

1.1.2 Ferromagentism

In ferromagnets the microscopic change at the phase transition is a loss of time reversal symmetry leading to atomic magnetic moments that orient in one distinct direction. Macroscopically domains appear and each exhibit a spontaneous magnetization, while to minimize the free energy of the crystal (or grain), the sum of the magnetic fields of these domains adds to a zero, unless an external field forces or poles the magnetic domains to align in one direction, resulting in a magnetic susceptibility over the crystals that is $\chi \gg 0$ and

$$\chi_{ij} = \frac{\delta M_i}{\delta H_j}, \qquad (1.1)$$

where H is the magnetizing force and M is the magnetization. Ferromagnetism only exists below the transition temperature T_C. Above T_C spins are disordered forming a paramagnet that does not show spontaneous magnetization. The temperature dependence of a paramagnetic spins follows the Curie law

$$\chi_m = \frac{C}{\theta - T} \qquad (1.2)$$

with Θ is the Curie temperature and C the Curie constant

$$C = \frac{Ng^2 \mu_B S(S+1)}{3k_B}, \qquad (1.3)$$

where S is the magnetic moment, N is the number of magnetic moments, g is the Landé g-factor, μ_B is the Bohr magneton and k_B is Boltzmann's constant. In a perfect paramagnet θ is zero while a paramagnet with a ferromagnetic transition θ is greater than 0 K and equates to T_C.

Louis Neéel in the 1950s predicted another class of magnets, called antiferromagnets where the lattice magnetization is zero due to the antiparallel alignment of the spin. Although many including Landau thought that this will not be possible as quantum fluctuation will annihilate the antiferromagnetic order, neutron diffraction soon afterwards showed that an enormous amount of materials exhibited antiferromagnetic order.

Between those two states (ferromagnetic and antiferromagnetic) a broad variety of other possible magnetic orders exist. Ferrimagnetic materials show spontaneous magnetization like a ferromagnet, but a part of the spins point in the opposite direction as in antiferromagnetic materials. The spins may also be canted and an antiferromagnet, may exhibit a ferromagnetic component orthogonal to the easy axis as shown in Figure 1.2. Another possible way to order is a proceeding rotation of the spins along all possible directions resulting in complex structures such as spin density waves, cycloids or helices.

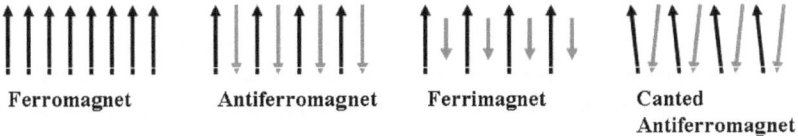

Ferromagnet **Antiferromagnet** **Ferrimagnet** **Canted Antiferromagnet**

Figure 1.2: *An illustration of spin order in ferromagnets, antiferromagnets, ferrimagnets and canted antiferromagnets*

1.1.3 Ferroelectricity

Only 11 out of 32 crystal classes that exist are polar, although 21 do not have an inversion center[3]. In these polar crystals a spontaneous electric polarization P_s[1] can exist without an external electric field. Those crystals are called pyroelectrics as P_s changes with temperature and vanishes at a transition temperature T_C. Another group of materials that show spontaneous polarization are ferroelectrics[4].

In ferroelectrics the microscopic origin at T_C is the loss of inversion symmetry where atomic displacement in one distinct direction produce an electric dipole. This may result in domains exhibiting a spontaneous polarization P_s in a macroscopic scale, also here reducing the free energy by compensating the polarization in all domains to zero in the bulk. Above T_C the ferroelectric material shows a paraelectric state that is comparable to the paramagnetic state.

---
[1]The naming of the polarization in the following text will refer explicitly to the electric polarization

If P_s can only appear in one axis (with two possible orientations) it is called a uniaxial ferroelectric, in contrast to a multiple axis ferroelectric. The dielectric permittivity ϵ of the paraelectric phase follows the Curie-Weiss-law for the dielectric constant

$$\epsilon = \epsilon_\infty + \frac{K}{T - T_0} \qquad (1.4)$$

with T_0 for the Curie temperature, K for the Curie constant and ϵ_∞ for the optical dielectric constant. T_C and T_0 are close to each other or at second order phase transitions are equal leading to an increased ϵ_∞ around T_0, often with higher values even in the ferroelectric phase. In the temperature regime between T_C and T_0 the paraelectric phase coexists with the ferroelectric phase with the paraelectric phase being metastable.

If the orientation of P_s can be reversed with electrical fields a hysteresis can be observed as shown in Figure 1.1. This separates ferroelectrics from pyroelectrics.

Usually ferroelectric crystals can be divided in two groups: order-disorder and displacive. The first one has a dipole in moment in each unit cell that are disordered at high temperatures, pointing in random directions. At the transition temperature the dipoles order, all pointing in the same direction within a domain. An example is $NaNO_2$. Displacive ferroelectrics show at the transition temperature a displacement of an ion against an other in the unit cell. This leads to an asymmetrical shift in the equilibrium ion positions and therefore to a permanent dipole moment. An example of such transition is $BaTiO_3$, where the Ti ion is displaced from the center of the TiO_6 octahedron at T_C.

Another type of ferroelectrics are known as relaxor ferroelectrics[5]. Relaxor ferroelectrics show characteristic properties, such as strong frequency dependence in the dielectric permittivity ϵ' as well as in the dielectric loss ϵ'' for temperatures at the peak maximum resulting in higher transition temperatures at higher frequencies. The phase transition is therefore uncertain and the crystal maintains a ferroelectric polarization into the paraelectric phase on warming.

1.2 Magnetoelectric Multiferroics

Magnetoelectric multiferroics are improper secondary ferroics where the ferroelectric and the ferromagnetic properties are coupled. While such behavior is predicted from symmetry considerations [6, 3, 7], how to actually create strong coupling between different types of ferroic order is a long standing issue in modern condensed matter physics [8, 9, 10, 11]. Early scientific work on this subject tried to combine ferroelectricity and magnetism in one material[12]. These two contrasting order parameters turned out to be mutually exclusive[13, 14, 15, 16, 17]. Further the coupling between the two is not strong in every case as the respective microscopic mechanisms are quite different and avoid strong interference[18, 19].

1.2. Magnetoelectric Multiferroics

There are certainly numerous technological applications for magnets and ferroelectrics, so to strongly couple them in a single material has been the aim of technological as well as scientific pursuit. While linear magneto-electric effect was realized in the 1960's[20], the work that followed failed to deliver materials that exhibited a strong magneto-electric coupling[8, 15]. For example ferroelectric magnets such as $BaMnF_4$, where the ferroelectric and magnetic transitions are separated by hundreds of degree Kelvin, showed little or no evidence of magneto-electric coupling[21]. More recent examples are $LuFe_2O_4$ or hexagonal $HoMnO_3$[22]. In $LuFe_2O_4$ at high temperatures, the Fe-ion is in a charge disproportionated state with a valence of +2.5[23]. Below T_{FE} = 330 K, e_g- charge order takes place that forms a layered arrangement of Fe^{2+} and Fe^{3+} ions with the consequence of inducing a ferroelectric distortion in the crystal. As the order parameter here is that of the charge ordering in Fe-ions and the associated lattice displacements, ferroelectricity is of an improper type. The antiferromagnetic order of Fe-ions occurs below T_N = 240 K and therefore is not coupled to the ferroelectric order. Also in the hexagonal $HoMnO_3$ a structural transition at T_{FE} = 875 K leads to the Mn^{3+} ions to be shifted in the position relative to the oxygen anions in two opposite directions, breaking inversion symmetry and inducing a dipole[24]. The antiferromagnetic transition occurs at a significantly lower temperature T_N = 75 K is clearly is not coupled to T_{FE}, however there is a magneto-electric effect reported to arise from the Ho ions[25].

The challenge has been how to devise materials where magnetism and ferroelectricity is strongly coupled. While much effort has been devoted to this area[26], very recently it was realized that frustration in the magnetic degrees of freedom in certain oxides can naturally couple magnetism and the lattice and result in strong magneto-electric effects[27]. Systems with competing magnetic interactions show many types of novel magnetic ordering. The result is a set of materials with spectacular properties such as giant magneto-capacitance[28], magnetically induced electric polarization flops[29, 18], and a rotation of electric polarization in a periodically varying magnetic field[30, 31]. The most important spin frustrated oxides that show strong magneto electric effects are the manganite perovskites with general formula $RMnO_3$ (R is a lanthanide or Y), the perovskite related RMn_2O_5, the delafaucites $MCoO_2$ and $MnWO_4$[27]. In this work the focus shall be on the perovskite manganites which are described in the following sections.

1.2.1 The Perovskite Structure

The Perovskite structure has three distinct atomic positions that can be described by the general chemical formula ABX_3, where A and B describing cations and X an anion. The X anion has approximately the same ionic diameter as the A cation forming a cubic dense sphere packing. The coordination of the B to the X ions is octahedral and the A cation occupies the space between these octahedra as shown in Figure 1.3. The size of the ions

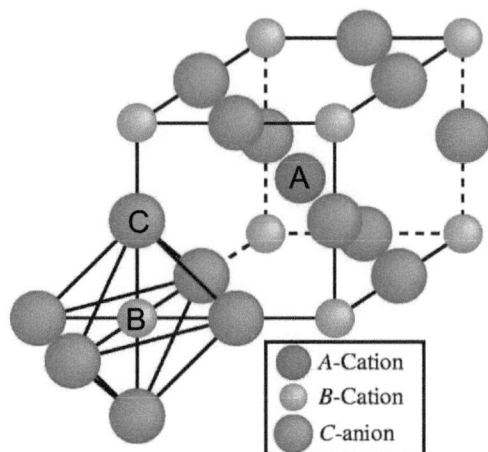

Figure 1.3: *An illustration of the crystal structure of cubic perovskite. The A-site cations are on the cube center position, the B-site cations are located on the corners of the cube and anions (here named C) are located on the edge centered position. The BX_6 octahedron is visible.*

plays a central role for the stability of the crystal system.

The variation of the ionic radii causes structural changes in perovskites[32]. Here the crystal structure can be predicted by the ratio of the ionic sizes. To say whether the ilmenite (hexagonal) or the perovskite (cubic/ rhombohedral/ orthorhombic) structure is more stable Goldschmidt proposed the formula

$$t = \frac{<r_A> + <r_X>>>}{\sqrt{2}(<r_A> + <r_X>)} \qquad (1.5)$$

where $<r_A>$ is the mean radii of the A ion, $<r_B>$ is the mean radii of the B ion and $<r_X>$ is the radii of the anion while t is the tolerance factor. Decreasing the size of the A-site atom leads to a decreasing tolerance factor, so for $RMnO_3$ at $R =$ La-Dy the orthorhombic system is preferred while for $R =$ Ho the hexagonal system is stable. A lower tolerance factor yields further to tilts between B-site octahedra and, therefore to modulations of the Mn-O-Mn bond angles. This is an important feature of these perovskite manganites as it shall be seen below the magnetic exchange between Mn ions can be tuned via the Mn-O-Mn bond angles.

1.2.2 Cooperative Jahn-Teller Distortion

With the electron configuration of [Ar] $3d^5$ $4s^2$ manganese has seven electrons in the outer shell. Trivalent Manganese has the configuration [Ar] $3d^4$, tetravalent [Ar] $3d^3$. In the latter case and considering Mn in an octahedral coordination, all three electrons are placed in t_{2g} orbitals. Following HundŠs law the first degenerated orbitals are filled with one electron in each d_{xy}, d_{xz} and d_{yz} orbital.

If there is an additional electron (as in trivalent manganese) this electron can be placed in a t_{2g} orbital to pair with another one (low spin configuration), but this needs a lot of pairing energy. Alternatively it can be placed in an e_g orbital (high spin configuration) which is energetically more favorable, i.e. $d_{z^2-r^2}$ or $d_{x^2-y^2}$. Since a single electron is placed in a degenerated state, the system acts as to break this degeneracy via a structural distortion and thus Mn^{3+} ions are Jahn-Teller active.

In terms of perovskites containing Mn^{3+} the Jahn-Teller structural distortion can take two forms. One form is that e_g electrons occupies a filled d_{z^2}-orbital, then the octahedron is elongated in the z-direction that gives two long bonds in z-direction and four short bonds in x,y-direction. In the case where the $d_{x^2-y^2}$-orbital is occupied, the octahedron is shortened in the z-direction with two short bonds and four long bonds in x,y-direction.

In RMnO$_3$, Mn has a trivalent Jahn-Teller active state and the $d_{z^2-r^2}$ orbitals order in a cooperative manner. This leads to a periodical ordering in the x-y plane, with long and short Mn-O bonding and an unoccupied e_g orbital perpendicular to that plane pointing along the z-direction[33] as shown in Figure 1.4.

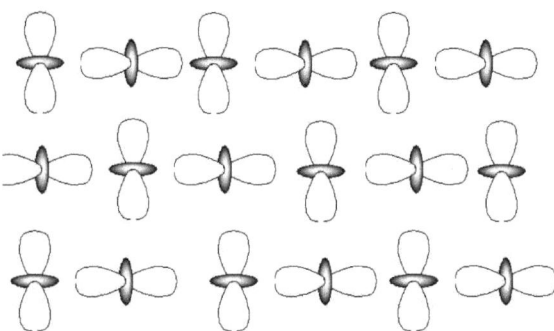

Figure 1.4: *Periodical change of the direction of the $d_{z^2-r^2}$ orbitals, causing a prolonged octahedra in this direction, resulting in the cooperative Jahn-Teller distortion as it happens in* LaMnO$_3$.

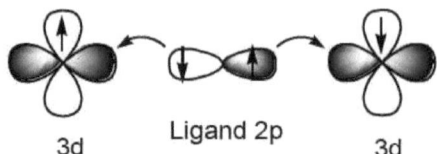

Figure 1.5: *The paired electrons of the 2p orbital interacts with the unpaired electrons in 3d of the Mn ions to the left and to the right resulting in a antiparallel alignment of Mn spins*

1.2.3 Magnetic Exchange in RMnO$_3$

Magnetic order in the RMnO$_3$ perovskite is determined essentially by super- and double exchange interactions. Both of these exchange interaction are mediated through the ligand O-anion and are in contrast to direct exchange that arises from the overlap of the electronic states of two adjacent magnetic atoms.

Super-exchange is an indirect, coupling of magnetic moments between next-nearest neighboring metals via an intermediate ligand-anion. The electron spin of an occupied metal d-orbital induces a parallel spin polarization in the neighboring anions fully occupied p-orbital. As this orbital contains two electrons the next-nearest metal ion will also induce a parallel coupling between the electron spin of its occupied d−orbital with the electron of the anions p-orbital. However due to Hund's rule, the electrons spins on the anion must be antiparallel and there the magnetic coupling between the two metal ions is antiferromagnetic.

In the case where one of the metal orbitals in this linkage has an unoccupied d-orbital, then the coupling between this unoccupied orbital and the p-orbital of the anion is ferromagnetic. This provides for a ferromagnetic coupling therefore between the two metal cations.

These models of magnetic exchange were first formulated by Goodenough[34, 35] and subsequently provided more rigorous mathematical underpinning by Kanamori[36]. They can be applied on interatomic spin-spin interactions between two metal atoms, each carrying a net spin, that interact by virtual electron transfers between the atoms and/or between a shared anion (like oxygen) acting as a ligand between the two metal atoms (super exchange).

A virtual electron transfer occurs between overlapping orbitals of electronic states that are separated by an energy ΔE. Orthogonal orbitals do not overlap, so there is no electron transfer and the exchange interaction between spins in orthogonal orbitals is a potential ferromagnetic exchange.

In the simplest form the rules set up by Goodenough and Kanamori state that super-exchange interactions are antiferromagnetic when the virtual electron transfer is between

1.2. Magnetoelectric Multiferroics

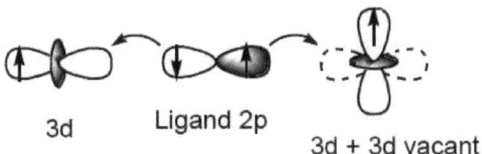

Figure 1.6: *The paired electrons of the 2p orbital interacts with the unpaired electron of the Mn ion to he left causing an antiparallel alignment of the spins of adjacent orbitals. However the interaction with the Mn on the right differs due to the perpendicular orientation of the d_{z^2}. This interactions between a filled O 2p state and an empty Mn 3d state resulting in a ferromagnetic alignment with the adjacent orbital and thus with the adjacent Mn ion.*

overlapping orbitals of the metal-ions that are each half-filled, but they are ferromagnetic when the virtual electron transfer is from a half-filled to an empty orbital or from a filled to a half-filled orbital.

In a case such as the manganite perovskites where the two cation orbitals overlap the same p orbital of a shared anion forming a cation-anion-cation arrangement with an angle close to 180°, it is helpful to introduce the virtual electron transfer from the shared anion to the cation first as the covalent component of the cation orbital. The net spin of the cation orbital is not changed by addition of a covalent component. However the covalent component extends the cation wavefunction over the anions to give an orbital overlap for the super exchange electron transfer, but a pure semicovalent antiferromagnetic exchange can occur between two empty orbitals provided each cation carries a net spin and the empty orbitals share the same anion p orbital. Figures 1.5 and 1.6 demonstrates ways of super exchange interaction between two anions over an cation resulting in antiferromagentic and ferromagnetic alignment.

In LaMnO$_3$ the Mn^{3+} has degenerated e_g-orbitals. This degeneracy is lifted at an rhombohedral to orthorhombic structural phase transition at high temperature[37]. This phase transition can be regarded as an orbital ordering transition where $3d_{x^2}$ and $3d_{y^2}$ orbital are occupied on alternate Mn sites within the orthorhombic ab-plane as shown in Figure 1.4. Structurally this leads to linkage in the lattice of alternating short and long Mn-O bonds. This orbital ordering causes half-filled d-bond orbitals to overlap empty d-bond orbitals in the ab-plane and as argued above such arrangements give rise to ferromagnetic interactions. Indeed below T$_N$ = 145 K[38] LaMnO$_3$ develops an antiferromagnetic order consisting of ferromagnetic layers in the ab-plane that are antiferomagnetically coupled along the c-axis. The antiferromagnetic stacking along the c-axis is understood also in terms of super-exchange as in this case the Mn exchange is via two unoccupied e_g orbitals. This magnetic order of ferromagnetic layers stacked antiferromagnetically is known as A-type[39].

Magnetic Frustration in RMnO$_3$ perovskites

By lowering the tolerance factor and decreasing the <Mn-O-Mn> bond-angles it was found that the Néel-temperature T_N decreases from 140 K for R=La to 45 K for R=Tb and increases again to 49 K R=Ho, as shown in Figure 1.7. This may indicate that the lower tolerance factor destabilizes and frustrates the A-type magnetic ordering. For R = Eu to Dy at T_N an incommensurate collinear spin density wave forms with Mn spin oriented along the b-axis. The magnetic wavevector for this order is $\kappa = qb^*$, and q varies between 0.2 to 0.39.

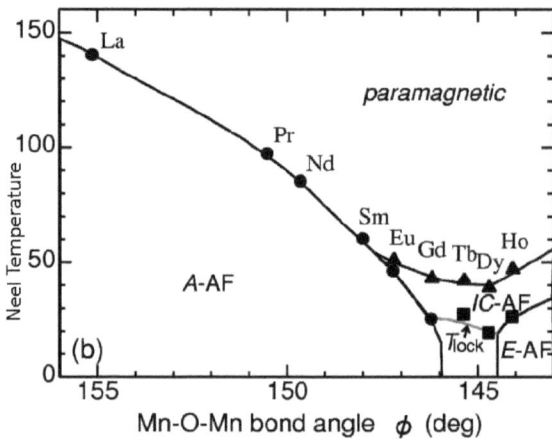

Figure 1.7: *The dependence of the Néel temperature as a function of <Mn-O-Mn> bond angle. The <Mn-O-Mn> bond angle is varied by choosing progressively smaller rare earth ions from La to Ho. Here it is evident that the A-type antiferromagnetic phase is suppressed with decreasing the size of the rare earth ions. For Eu, Gd, Tb and Dy, an incommensurate magnetic structure is evident at lower temperatures. For Ho a commensurate E-type structure is found. This Figure was published by Kimura et al.[40]*

In contrast compositions that exhibit only A-type antiferromagnetic order, the ones forming a incommensurate spin order below T_N (R = Eu to Ho) show an additional transition below T_N. For R = Eu and Gd the system stabilizes in an A-type antiferromagnetic structure again, while at R = Ho, Tm, Yb, Lu the system stabilizes in a commensurate antiferromagnetic structure known as E-type[41, 42, 43, 44]. Here in the E-type order one of the four Mn ions in the orthorhombic unit cell points antiparallel to the other three while the antiferromagnetic stacking along the c-axis is maintained. For R = Tb and Dy a second component of the magnetic develops in addition to the b−axis component, a phase shift difference of approximately $\frac{\pi}{2}$ leading to a cycloidal (also referred to as spiral) order of the spins below T_s causing a spontaneous polarization[45, 46, 47].

To understand the origin of the cycloidal multiferroic manganites RMnO$_3$ leading to a spontaneous polarization a model based on two e_g-orbitals double-exchange model can be used. Three factors play a role of determining the magnetic order: The nearest-neighbor (NN) super-exchange, the next-nearest-neighbor (NNN) super-exchange and the Jahn-Teller lattice distortion. The arrangement of the couplings in displayed in Figure 1.8

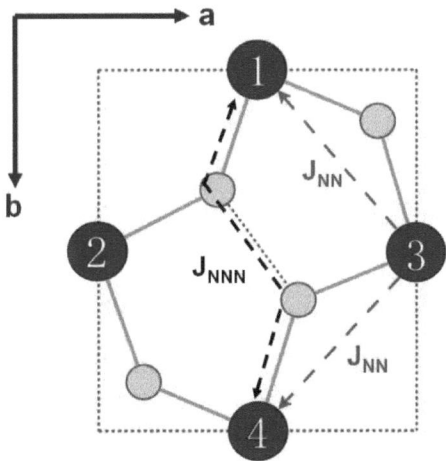

Figure 1.8: *Sketch of the crystal structure in ab-plane of RMnO$_3$ with the NN super exchange coupling (J_{NN}) and NNN super-exchange coupling (J_{NNN}). Filled spheres with numbers represent the Mn ions while the light spheres represent the oxygen ions. This Figure was published by Dong et al.[48].*

The NN coupling is the strongest coupling of those in this system and can be written as follows:

$$\mathbf{H}_{DE+SE} = -\sum_{<ij>}^{\alpha\beta} t_r^{\alpha\beta} \Omega_{ij} \tilde{c}_{i\alpha} c_{j\beta} + J_{AF} \sum_{<ij>} S_i \cdot S_j \quad (1.6)$$

where the first term is the double exchange kinetic energy of the e_g electrons, the second term is the NN super exchange[2]. The double exchange hopping amplitude $t_r^{\alpha\beta}$ is orbital- and direction-dependent. Calculations indicate that the total energy of the spiral phase is

---
[2]Although the magnetic interaction is mainly based on super exchange the double exchange must be included in the Hamiltonian for a better result[48].

always higher than the energies of the A-type magnetic structure, regardless of the value of J_{AF}. So the spiral phase can not be the ground state.

The NNN coupling is much weaker as the NN coupling as it runs over two anions (approximately 10% of the NN superexchange). Further the coupling is anisotropic as $b > a$ and the coupling between Mn(1)-O-O-Mn(4) (J_{NNN1}) is 2.2 times higher as in Mn(2)-O-O-Mn(3) (J_{NNN2}). Additionally this NNN coupling is antiferromagnetic leading to an E-type system as in HoMnO$_3$ when the coupling is strong enough. The Hamiltonian reads

$$\mathbf{H}_{J_{NNN}} = \sum_{[ij]} J_{2\gamma} S_i \cdot S_j \qquad (1.7)$$

where γ runs over a and b.

So if the <Mn-O-Mn> angles are changed by varying the tolerance factor, the NNN coupling frustrates the system leading to the observed incommensurate spin spiral, but with lower q values as observed experimentally. Taking the Jahn-Teller lattice distortion into account that prolongs or shortens the Mn-O bonds in the Mn-O-Mn path realistic values can now be calculated.

The development of the second component of the magnetic order in the compositions TbMnO$_3$ and DyMnO$_3$ though do not only indicate the onset of the spin cycloid but also the onset of a spontaneous ferroelectric polarization P_s along the c−axis[27, 19].

It was shown phenomenologically [49] that the onset of ferroelectricity can induced a ferroelectric polarization if the the magnetization M is inhomogeneous over the crystal. This arises from Maxwell's equations which allow for a third order term of $PM\partial M$. The symmetry of a cubic crystal constrains the possible forms of a magnetically induced polarization and can be written as[50, 49, 51, 52]

$$P \propto [(M \cdot \partial)M - M(\partial \cdot M)] . \qquad (1.8)$$

The magnetic spin cycloidal structure works well with this formulation as it breaks not only time reversal but also with space inversion symmetry[53]. This is because the cycloidal structure is chiral and does not have an inversion center. To calculate the polarization induced by magnetization it is necessary to describe the magnetic order precisely. A magnetic cycloidal structure can be described generally as a sine wave with two orthogonal components, M_x and M_y, propagating in the same direction with the propagation vector k that have a phase shift of $\frac{\pi}{2}$ and a constant component M_z parallel to the unit vector. This can be described by the expression[27]

$$M = M_x sin(2\pi k \cdot x) + M_y cos(2\pi k \cdot x) + M_z . \qquad (1.9)$$

1.2. Magnetoelectric Multiferroics

By setting the direction of the propagation vector k along x or y a cycloidal structure is described, while if k points along z, the formula describes a helical structure. When M_z is not zero a conical component is added. Due to the fact that adjacent spins prefer to align in a collinear manner the non-collinear cycloid must be stabilized by the antisymmetric Dzyaloshinskii-Moriya interaction[54, 55] [3]

$$\mathbf{H}_{DM} \propto D_{ij} \cdot [S_i \times S_j] \qquad (1.10)$$

where D_{ij} is the Dzyaloshinskii-Moriya vector relating the spins S_i and S_j. As the spins are connected via cross product the amplitude of this interaction favors non-collinear spin arrangements. To calculate the amplitude and direction of the polarization arising from the cycloidal order Equation (1.9) can be substituted in Equation (1.8) leading to

$$\bar{P} = \frac{1}{V}\int d^3x P = \gamma\chi_e M_x M_y \left[(S_i \times S_j) \times k\right] \qquad (1.11)$$

where γ in a coupling constant, χ_e is the electric susceptibility and S_i and S_j are two adjacent spins. This equation sets up conditions to the polarization induced by the magnetic structure. First, of course, the parameters outside the squared bracket must be non-zero, what is the case in a spiral but not for a collinear spin density wave. Secondly the adjacent spins must not be collinear, what is the case in a spiral. Third the unit vector of the spiral and the direction of its propagation also must not be parallel, what is the case in a cycloid but not in a helix.

These conditions can be realized in the transition of $TbMnO_3$ and $DyMnO_3$ to cycloidal order. Due to magnetic anisotropies in the crystal the first component of the magnetic order is at T_N along the easy axis forming a spin density wave. The second component orders at T_s what then forms the cycloidal structure and only than a ferroelectric polarization is induced[56, 57, 58, 49].

The mechanism of how a magnetic spin cycloid breaks inversion symmetry and induces a ferroelectric polarization in a crystal is still uncertain although two opposing theories are known. The first is based on the Dzyaloshinskii-Moriya interaction. This interaction favors orthogonal spin order and is proportional to the spin-orbit coupling constant λ. This constant is dependent on the position of the oxygen ion between two magnetic transition metal ions. In a spin cycloid the system maximizes the energy of the Dzyaloshinskii-Moriya interaction by pushing negative oxygen ions in one direction perpendicular to the spin chain formed by magnetic ions and though inducing an electric polarization perpendicular to the chain[45, 59, 60]. Another theory uses the idea of a spin current, $j_{i,j} \propto S_i \times S_j$ describing the precession of the spin S_i in th exchange field created by the spin S_j. The induced electric dipole in then given by $P_{i,j} \propto r_{i,j} \times j_{i,j}$[50].

---
[3]The Dzyaloshinskii-Moriya interaction is a relativistic correction to super-exchange and its strength is proportional to the spin-orbit coupling

1.3 Motivation and Scope of This Work

Although the properties of the multiferroics $TbMnO_3$ and $DyMnO_3$ are known to a great extend, the transition from collinear A-type antiferromagnetic order to the cycloidal structure is still unclear and the topic of recent studies[61, 62, 63, 64]. The R elements at the threshold, that is Eu, Sm and Gd, are highly neutron absorbing and structural and magnetic analysis with neutron diffraction is not possible or only under very special conditions, for example taking using expensive isotopic samples. The goal is now to imitate the structural and magnetic properties of the compositions at the transition zone by replacing the rare earth R with elements that can be investigated with neutrons. With this method the stages of the transition from collinear to cycloid can be analyzed, like is it of first or second order or what is the dependence of the ferroelectric polarization. In this work Nd in $NdMnO_3$ is replaced by Y to reduce the tolerance factor and therefore to simulate the magneto-electric properties of $RMnO_3$.

In literature no information is published about a series where Nd in $NdMnO_3$ is replaced by Y. Nevertheless other comparable compounds were synthesized, showing that those properties can indeed be simulated and giving some information about the possible synthesis conditions, $Eu_{1-x}Y_xMnO_3$[65, 66, 67] and $Gd_{1-x}Tb_xMnO_3$ [68] are examples for this. $Nd_{1-x}Y_xMnO_3$ covers the range of rare earth manganites from orthorhombic $NdMnO_3$ ($x = 0$) to $YMnO_3$ ($x = 1$)[4].

This thesis shall show that it is possible to grow high quality single crystals of the $Nd_{1-x}Y_xMnO_3$ solid solution between the composition range of $0 < x < 0.55$, while for $x > 0.55$ the orthorhombic perovskite structure is no longer stable. In chapter 3 describes the synthesis of powder and single crystals samples of this solid solution, while chapter 4 describes a characterization of the general $Nd_{1-x}Y_xMnO_3$ magnetic and structural phase digram using powder samples. In chapter 5 the results of the physical properties measurements of the single crystals are displayed while in in chapter 6 the temperature dependent neutron diffraction measurements of the single crystals are shown. Chapter 7 describes the detailed structure of the magnetic order at 2 K and 30 K of the samples $x = 0.45$ and 0.55 are while in chapter 8 the results of this work are summarized and discussed.

---

[4]Y is not a rare earth element, but as Ho has the same ionic diameter as Y, $YMnO_3$ and $HoMnO_3$ both are structurally alike[69] and can be used in a equivalent manner in synthesis. Of course the magnetic contribution of Ho and Y in this crystal is completely different

Chapter 2
Experimental Methods

In this chapter different methods will be presented that were used to examine the polycrystalline and the single crystal samples and to investigate the crystallographic, nuclear and magnetic structure and phase transitions of $Nd_{1-x}Y_xMnO_3$.

2.1 X-ray Diffraction

X-rays are electromagnetic radiation of wavelength about 1 Å, which is about the same size as an atom. The discovery of X-rays in 1895 by Wilhelm Conrad Röntgen enabled scientists to probe crystalline structure at the atomic level. X-ray diffraction has been in use in two main areas, for the fingerprint characterization of crystalline materials and the determination of their structure. Each crystalline solid has its unique characteristic X-ray powder pattern which may be used as a 'fingerprint' for its identification. Once the material has been identified, X-ray crystallography may be used to determine its structure, i.e. how the atoms pack together in the crystalline state and what the interatomic distance and angle are. X-ray diffraction is one of the most important characterization tools used in solid state chemistry and materials science.

The wavelength of the x-ray beam and the distance between the scattering planes in the crystal leads to a distinct scattering angle θ and is described by Bragg's law

$$n\lambda = 2dsin\theta \tag{2.1}$$

with the layer distance d, wave angle θ and wavelength λ. The optical retardation of the reflections on the scattering planed induces destructive or constructive interference. If the optical retardation is a multiple of the wavelength, the reflected rays will be amplified at most.

Peak intensity I of diffracted x-rays on a crystal is described by

$$I = \left|G^2\right|\left|F(Q,T)^2\right|PLE \tag{2.2}$$

where G is a lattice vector, F for the form factor, P for the polarization factor and L for the Lorentz factor. E describes the possible errors due to the measurement. The most important variable for structural refinement is the structure factor that is proportional to the intensity

$$I_{hkl} \propto \left|F_{hkl}^2\right|. \tag{2.3}$$

The structure factor F_{hkl} is obtained by a summation of all atoms in the unit cell. For x-rays the term f_i is the form factor and varies with $\sin\theta$ over λ, while for neutrons this is replaced with a scalar value of b which is the neutron scattering length.

$$F_{hkl} = \sum_j^N f_i \, e^{2\pi i(hx_j+ky_j+lz_j)} \, e^{-W_j(t)} \tag{2.4}$$

The relative positions of the atoms in the unit cell are xyz and hkl are MillerŠs indexes. The form factor f is equal to the Fourier transformation of the electronic density (at x-ray diffraction) of an atom that is mostly regarded as spherical. $W_j(T)$ is the Debye-Waller factor considering the thermal movement of the diffracting atoms.

Full width at half maximum (FWHM) and shape of the peaks in diffraction patterns are influenced by the measurement parameters and sample properties. Measurement parameters are the spectrum of the beam, beam divergence, beam focus, grid width and resolution. To get comparable diffraction pattern and reliable results it is very important to keep these parameters constant. Affecting sample properties are sample volume for transmission spectrometer or sample area for refection spectrometer as well as size, structure and dispersion of the powder grains.

X-rays are electromagnetic radiation that interacts with the electron shell of the atoms, i.e. the more electrons an atom or ion has, the stronger the reflections. Other properties of the sample than electron density, e.g. atomic charge or magnetic orientation, play no role for laboratory, non polarized x-rays[1]

X-rays with energies from 10 to 50 keV can be used for atomic structure determination as it is described above. In this work the diffraction patterns were made with a Bragg-Brentano diffractometer as it is described in Section A.10.

The disadvantage of x-ray measurements is the fact that the intensities of the measured reflections are dependent of the electron density. That leads to the fact that when a crystal has heavy atoms like rare earth elements their reflections are so strong that relatively light elements like hydrogen or oxygen are hardly be seen in the pattern. Only the form factor can give a hint for very sharp long range measurements.

Regarding $Nd_{1-x}Y_xMnO_3$ oxygen may not be considered in the Rietveld refinements, but the ratio of Nd, Y and Mn to each other (with one fixed) is accessible. So setting Mn

---

[1] As a electromagnetic wave the magnetic part of the x-ray beam also interacts with the magnetic part of a structure, but the intensity of the scattered magnetic pattern is very low (approximately to the factor 10^{-6} less) and therefore can only be detected with high resolution instruments like in synchrotron experiments using polarized x-rays.

to 1, the right stoichiometry of Nd and Y can be calculated from the diffraction pattern. This is especially useful as the coherent neutron scattering cross section for Nd and Y are close to equal: $\sigma_{coh}(Nd) = 7.43$, $\sigma_{coh}(Y) = 7.55$. This makes it very hard to distinguish the ratio of Nd and Y via neutron scattering.

2.1.1 Laue diffraction

To determine the quality of a single crystal and to orient the crystal for single crystal scattering experiments, Laue diffraction was used. Here a white x-ray (or neutron) beam is directed onto a crystal. Following Bragg's law (see Equation (2.1)) for a interatomic spacing distance d there is a λ that scatters at a certain Θ. As this is only valid on a certain angle in space set by the orientation of the crystal the scattering will show spots where these conditions are fulfilled. With this Laue image the orientation of the crystal in space can be calculated and the quality of the crystal can be assessed. This is done by examining the spherical nature of the diffraction spots and checking to see if they can be indexed using a single crystal orientation. Twinned or multiple grain crystals will produce multiple diffraction spots centered at the expected position of a bragg peak from a single crystal. Powder samples will produce rings.

2.2 The Rietveld method

To analyze powder diffraction patterns, H.M. Rietveld published ideas for mathematical refinement[70, 71] of diffraction patterns that could be used efficiently after powerful computers made their way in scientific analysis. His method refines a crystallographic model by minimizing the difference of the the points of the measured diffraction pattern and the calculated diffraction pattern with the method of the least squares.

$$\chi^2 = \sum_i \frac{|obs_i^2 - calc_i^2|}{obs_i^2} \qquad (2.5)$$

The data measured in step scan mode has distinct values for distinct measurement angles φ_i. It is necessary to have a model for the measured structure in advance. Starting from that structural model theoretical values for the angles φ_i can be calculated. Then the measured and calculated intensities on every angle φ_i will be compared and a quality of fit χ^2 can be calculated. On basis of that value the model can be optimized by varying the structural parameters as the size and symmetry of the unit cell as well as the reflection multiplicities and the peak shape functions. By successfully solving the structure factor (with a minimized χ^2) the atomic positions are known. To characterize the quality of the fit of one calculated phase to the observed diffraction, Bragg's R-Parameter can be computed.

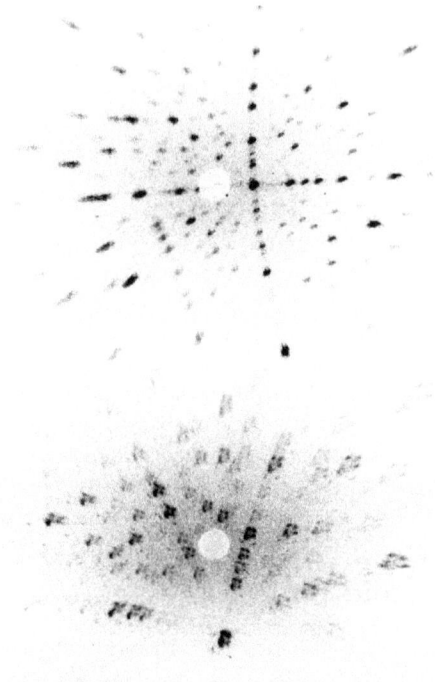

Figure 2.1: *Laue image of two crystals of $x = 0.4$ from different crystal growths taken on OrientExpress at the ILL. Top image shows no other visible grains and sharp spots, bottom image shows clearly several grains with orientations differing up to 5 degrees.*

$$R_B = \frac{\sum_i |I_{io} - I_{ic}|}{\sum_i I_{io}} \qquad (2.6)$$

2.3 Neutron Scattering

Neutron scattering is used in many scientific fields. A neuron beam can be produced by nuclear fission in a nuclear reactor or in a spallation source facility. A nuclear reactor produces a constant neutron flux whereas a spallation source produces a pulsed beam.

Neutron diffraction compared to x-ray or electron diffraction is not proportional to the number of electrons but a function of the energy levels in the nuclei. As the volume the

2.3. Neutron Scattering

nuclei is less than 10^{-12} of the whole volume the matter is highly penetrated by neutrons and surface phenomena are negligible. Since neutrons are not charged, they are not scattered by the charge of the particles, unlike x-ray and electron diffraction. Nevertheless the neutrons have a nuclear spin that is able not only to interact with other nuclear spins but also with electron spins. Ordered electron spins, like in ferromagnets and antiferromagnets, are the origin of magnetic neutron scattering (see next Section). Neutron diffraction reveals structural details of the target material, which are measured by recording the way in which neutrons are deflected (elastic scattering). Neutrons can also change their energy during the scattering experiment; this is used to study the types of vibrations that can occur in the solid (inelastic scattering). An important difference between neutron and X-ray diffraction is that neutrons are sensitive to aligned magnetic moments in the material.

Due to the wave-particle character of neutrons, expressed in the formula of de Broglie

$$\lambda = \frac{h}{p} = \frac{h}{mv} \tag{2.7}$$

the wavelength of the neutrons varies with the speed of the neutrons. Neutrons with a resulting wavelength of 5 to 0.6 nm or with the energy of 0.025 until 0.2 eV, this is equivalent to speeds from 2ů103 to 6ů103 $\frac{m}{s}$ are called thermal neutrons. They will be used in this work.

The positions of the reflections can be calculated with BraggŠs law analog to the x-ray patterns but the intensities of the peaks are completely different, of course. Inelastic neutron scattering can be used for measurement of correlated motions of atoms (coherent scattering) or measurement of self-correlations e.g. diffusion (incoherent scattering).

The neutron can interact with matter in two ways. First the strong interaction causes the neutron to scatter on the nucleus. Second the magnetic dipole of the neutron of μ_N = -1.9130428, where μ_N is the nuclear magneton, interacts with the magnetic moments in the crystal so that the magnetic structure can be investigated.

The magnetic neutron scattering shows an similar angular dependence as the structural scattering. The total scattering pattern F_{tot} (hkl) has therefore structural F_{core} (hkl) and magnetic F_{mag} (hkl) components:

$$|F_{tot}(hkl)|^2 = |F_{core}(hkl)|^2 + q^2 |F_{tot}(hkl)|^2 . \tag{2.8}$$

The factor q^2 is equal to $\sin^2(\alpha)$ with α being the angle between reflecting plane and magnetization vector. The unit cell constants of structural and magnetic parts may differ resulting in different scattering reflections.

The general ansatz for the waves is

$$\Phi_1 = e^{i\vec{k}\vec{r}} \tag{2.9}$$

with

$$k = |\vec{k}| = \frac{2\pi}{\lambda} \tag{2.10}$$

Though the scattered wave is

$$\Phi_2 = -\frac{b}{r} e^{i\vec{k}'\vec{r}} \tag{2.11}$$

with r being the distance to the scattering core, $\vec{k}$ and $\vec{k}'$ being the wave vector and b as the scattering length.

2.3.1 Magnetic Neutron Scattering

The magnetic diffraction potential based on the magnetic interaction of the neutron dipole moment with the magnetic field $\vec{H}_e$ of the electron is not radial symmetric and works over a long range.

$$V_j(\vec{xj}) = -\gamma\mu\vec{\sigma} \cdot \vec{H}_e \tag{2.12}$$

The magnetic scattering cross section is a result of the magnetic form factor as follows

$$\left(\frac{d^2\sigma}{d\Omega dE'}\right)_{\lambda \to \lambda'} = (\gamma r_0)^2 \frac{k'}{k} \sum_{\alpha\beta}(\delta_{\alpha\beta} - \hat{Q}_\alpha \hat{Q}_\beta) \cdot \sum_{l'd'}\sum ldF_{d'}^*(\vec{Q})F_{d'}(\vec{Q})\cdot$$
$$\cdot \sum_{\lambda'\lambda} p_\lambda \left\langle \lambda | exp(-i\vec{Q}\cdot\vec{R}_{l'd'})S_{l'd'}^\alpha | \lambda' \right\rangle \cdot \left\langle \lambda' | exp(i\vec{Q}\cdot\vec{R}_{ld})S_{ld}^\beta | \lambda \right\rangle \cdot$$
$$\cdot \delta(E_\lambda - E_{\lambda'} + E - E'). \tag{2.13}$$

Here $\hat{Q} = \frac{\vec{Q}}{|\vec{Q}|}$ is the normalized scattering vector $|\gamma r_0| = 5.39 \cdot 10^{-15}$ m is the coupling constant, α and β are the indexes for the cartesian components and p_λ the probability to find the scattering system in the initial state λ. As a crystal is periodically ordered it is now possible to introduce an elementary cell that is shifted by the vector $\vec{l}$ to the origin. The magnetic ions in the cell can be numbered consecutively with d. The spin of the electrons is described with S that can be extended to a time dependent Fourier transformed spin operator

$$S_{ld}^\beta(t) = e^{i\mathcal{H}t/\hbar} S_{ld}^\beta e^{-i\mathcal{H}t/\hbar}. \tag{2.14}$$

The lattice vibrations will be neglected, so the atomic positions are time independent. Only the Debye-Waller-factor W_d for the magnetic diffraction cross section is considered, including the shift from the equilibrium u_d in the factor $W_d = \frac{1}{2}\left\langle (\vec{Q}\cdot\vec{u}_d)^2 \right\rangle$.

2.4. Magnetization Measurements

When the dynamic of the spin system with the diffraction law $\mathcal{S}^{\alpha\beta}_{dd'}(\vec{Q},\omega)$ and the other factors in $\mathcal{A}_{dd'}(\vec{k},\vec{k}')$ are summarized the result is

$$\left(\frac{d^2\sigma}{d\Omega dE'}\right)_{\lambda\to\lambda'} = \frac{k'}{k}\sum_{\alpha\beta}(\delta_{\alpha\beta}-\hat{Q}_\alpha\hat{Q}_\beta)\cdot\sum_{dd'}\mathcal{A}_{dd'}(\vec{k},\vec{k}')\cdot\mathcal{S}^{\alpha\beta}_{dd'}(\vec{Q},\omega) \quad (2.15)$$

with

$$\mathcal{A}_{dd'}(\vec{k},\vec{k}') = \frac{N}{\hbar}(r_0\gamma)^2\frac{k'}{k}F^*_{d'}(\vec{Q})F_d(\vec{Q})\ e^{-W_{d'}}e^{-W_d} \quad (2.16)$$

$$\mathcal{S}^{\alpha\beta}_{dd'}(\vec{Q},\omega) = \frac{1}{2\pi}\sum_l\int dt\ e^{(i[\vec{Q}(\vec{R}_{ld}-\vec{R}_{0d'})-\omega t])}\left\langle S^\alpha_{0d'}(0)S^\beta_{ld}(t)\right\rangle \quad (2.17)$$

Any disturbance of a magnetic system can be described with a changing magnetic field $\vec{H}(\vec{r},t)$ in time and space. The magnetization has a direct connection to the tensor of the susceptibility χ. The relaxation function stands in opposition to that operator reading as follows

$$\mathbf{R}^{\alpha\beta}(\vec{K},t) := \left\{S^\alpha(-\vec{K},0),S^\beta(\vec{K},t)\right\} \quad (2.18)$$

with the curly brace defined as a connection to

$$\{A,B\} := \int_0^{1/k_BT}\left\langle e^{\lambda\mathcal{H}}A\ e^{\lambda\mathcal{H}}B\right\rangle d\lambda - \frac{1}{k_bT}\langle A\rangle\langle B\rangle. \quad (2.19)$$

If the elastic coherent Bragg part of the scattering law $\mathcal{S}^{\alpha\beta}_{Bragg}(\vec{Q},\omega)$ is subtracted only the diffuse and inelastic scattering part is left considering N magnetic elementary cells.

$$\mathcal{S}^{\alpha\beta}-\mathcal{S}^{\alpha\beta}_{Bragg} = N^{-1}\frac{\hbar\omega\beta}{1-e^{-\hbar\omega\beta}}\mathbf{R}^{\alpha\beta}(\vec{Q},\omega) \quad (2.20)$$

$$= \frac{k_bT}{g^2\mu_B^2}\chi^{\alpha\beta}(\vec{Q})\frac{\hbar\omega\beta}{1-e^{\hbar\omega\beta}}\mathbf{F}^{\alpha\beta}(\vec{Q},\omega) \quad (2.21)$$

It is already the statistical approximation for χ that does not have a time dependent term any more. A frequency spectrum for the fluctuations $\mathbf{F}^{\alpha\beta}(\vec{Q},\omega)$ has been introduced.

2.4 Magnetization Measurements

The measurement of the magnetic properties is an essential step when magnetic materials are studied. The macroscopic magnetization M, as a sample property, depends in general on the magnetic field H and the temperature T as

$$M = \chi_m H, \quad (2.22)$$

with χ_m as the volume magnetic susceptibility that is a material constant at a constant temperature that was shown in formula 1.2 resulting in

$$M = \frac{C}{\theta - T} H. \qquad (2.23)$$

Depending on the crystal structure it can also show more dependencies, e.g. from pressure p or electric field E in materials showing the Villari-effect or a magneto electric effect, respectively. The measurements in this work are limited in to a measurement of $M(T,H)$.

To determine the magnetic properties of the samples a PPMS magnetometer was used with a VSM setup. The vibrational frequency of the sample in the VSM was about 40 Hz with an amplitude between 0.5 mm and 1 mm.

2.5 Dielectric Measurements

The investigation of polarization P and ϵ is necessary to understand the properties of ferroelectrics. P depends on the applied electric field E and the temperature T. In magnetoelectric materials an additional dependence on magnetic field H exists[2].

The polarization P_i ($\frac{C}{m^2}$) that is induced in an insulating polarizable material (a dielectric) by an applied electric field vector E_i ($\frac{V}{m}$) is given by

$$P_i = \chi_{ij} E_j \qquad (2.24)$$

where χ_{ij} ($\frac{F}{m}$) is the second-rank tensor known as the dielectric susceptibility of the material. Equation (2.24) is valid only for linear materials on in a linear limit for nonlinear materials and, in general, P_i depends on higher-order terms of the field. The total surface charge density is induced in the material by the applied field is given by the dielectric displacement vector $\mathbf{D}_i$ ($\frac{C}{m^2}$)

$$D_i = \epsilon_0 E_i + P_i \qquad (2.25)$$

with ϵ_0 as the dielectric permittivity of a vacuum. It follows from Equation (2.24) and Equation (2.25) that

$$D_i = \epsilon_0 E_i + \chi_{ij} E_j = \epsilon_0 \delta_{ij} E_j + \chi_{ij} E_j = \epsilon_{ij} E_j = k_{ij} \epsilon_0 E_j \qquad (2.26)$$

where ϵ_{ij} is the dielectric permittivity of the material and δ_{ij} is Kronecker's delta. For most ferroelectric materials $\epsilon_0 \delta_{ij} << \chi_{ij}$ and $\epsilon_{ij} = \chi_{ij}$. In practice, the relative dielectric

---
[2]Materials with further dependencies, e.g. on pressure, exist. As no such dependencies have been investigated for this work, this will not be discussed here.

permittivity, $k_{ij} = \frac{\epsilon_{ij}}{\epsilon_0}$ also known as the dielectric constant of the material, is more often used than the dielectric permittivity.

More details of this measurement is displayed in Section A.4.

2.6 Pyroelectric Measurements

To obtain the ferroelectric polarization P the pyroelectric current can be measured. The pyroelectric effect is defined as the change of the spontaneous polarization P_s as a function of temperature T[72]. The origin of this effect is that, when all ferroelectric domains are aligned in the same direction the polarization induces a charge separation that remains even when both sides are short cut. When the temperature of the sample is increased above phase transition temperature the polarization is quenched and the separated charges try to compensate. When both sides are connected, a charge flow can be measured that is directly connected to the strength of the polarization at ground state.

The pyroelectric current can be measured after cooling the sample, ideally in the shape of a plate, in an electric field and heating up without an external electric field in a typical heating rate of $\frac{dT}{dt} = 1 - 2 \frac{K}{min}$. The resulting pyroelectric current can be measured directly with an ampere meter connected to the surfaces of the sample. It is given as

$$\mathbf{I}(T) = A_{sample} \left(\frac{dT}{dt}\right) p \qquad (2.27)$$

where A_{sample} is the surface area of the sample and p is the pyroelectric coefficient

$$p = \frac{dP}{dT}. \qquad (2.28)$$

From the measurement of I(T) it is possible to calculate p and thus to obtain P by integration over T.

It is a common approach to measure both the dielectric constant ϵ and the polarization $P(T,E,H)$. Generally it is not possible to calculate P directly from ϵ, but occurring anomalies in ϵ can be evidence for changes in P, and thus conclusions about P can be made. As pointed out in Section 2.5, an often very large anomaly in ϵ exists especially at the paraelectric to ferroelectric transition.

Chapter 3
Synthesis

$Nd_{1-x}Y_xMnO_3$ is easily accessible with solid state reaction technique, although the temperature and oxygen partial pressure must be controlled to receive the right manganese valence[73, 74]. Changing the rare earth elements from Nd to Ho in $RMnO_3$ with R = Nd to Ho showed that except Dy and Ho all compositions were accessible with solid state reaction technique and could be grown as single crystal samples in a floating zone furnace. In Dy and Ho the thermodynamically stable crystal structure is hexagonal. However metastable orthorhombic $RMnO_3$ powder samples with R = Dy, Ho or Y were made with a solution based procedure via complexation by citrates[69].

A similar series that has been in the focus for some years now is $Eu_{1-x}Y_xMnO_3$. Also these compounds were made with solid state reaction techniques. The single crystal samples of these were prepared in a floating zone furnace as it will be described in Section 3.2.1under comparable conditions. Here was observed that with samples of $x \geq 0.75$ hexagonal impurities formed.

In this work first the polycrystalline samples of $Nd_{1-x}Y_xMnO_3$ were synthesized. This step served information about the changing preparation conditions with x and the maximum amount of Y for receiving single orthorhombic phase samples. At $Eu_{1-x}Y_xMnO_3$ the threshold was 0.75[67], for $Nd_{1-x}Y_xMnO_3$ the maximum doping was $x \geq 0.6$. Further with polycrystalline samples the crystal structure and general magnetic properties can be checked quickly and a first phase diagram may be mapped out.

For looking at the details of the magnetic structure and the transitions that takes place it is necessary to grow single crystals. Only with single crystals anisotropic physical measurements can be done and every single crystallographic reflection may be observed separately in the Ewald's sphere with neutron scattering experiments.

All samples made for this thesis were synthesized at the Helmholtz-Zentrum Berlin für Materialien und Energie using the chemical labs, box furnaces, tube furnaces and image furnaces of this facility. The following section describes the method of preparing the polycrystalline sample and of the growth of single crystals of $Nd_{1-x}Y_xMnO_3$.

3.1 Synthesis of Polycrystalline $Nd_{1-x}Y_xMnO_3$

Polycrystalline samples of $Nd_{1-x}Y_xMnO_3$ are important as they are used in Section 4 to initially map out the structural and magnetic transition with changing x. With powder diffraction data it is possible to measure the crystallographic reflections over a wide range of 2Θ with only one measurement and to receive information about impurities, other phases (like hexagonal phases), and, depending on x-rays or neutron diffraction, the stoichiometry of the atoms. Measuring temperature dependent magnetic susceptibility and neutron diffraction, also the transition temperatures can be determinated. Further polycrystalline samples are used as precursors for single crystal growth.

Temperature and atmosphere play a central role in forming the right crystallographic phase. Temperature influences the formation of either orthorhombic or hexagonal crystal structure[69] and also the stoichiometry of the manganese sites[75, 76] in the crystal[1] and of course the speed of the solid state reaction. Higher oxygen pressure stabilizes the orthorhombic phase[69] and also stabilizes the Mn^{3+} oxidation state[75, 76].

Samples of the solid solution $Nd_{1-x}Y_xMnO_3$ were prepared by solid state reaction chemistry. Stoichiometric precursors of Nd_2O_3, Y_2O_3 and Mn_2O_3 with purity >99.99% were balanced, mixed, thoroughly grinded and fired for various times at high temperatures. After balancing, mixing and grinding, the precursors were fired at 1200°C, 1280°C and 1350°C for 12 hours each with intermediate grinding.

Samples with $x = 0.0, 0.1, 0.2$ were prepared under nitrogen atmosphere to prevent oxidation of Mn^{3+} to Mn^{4+} [74]. Samples with $x = 0.3$ and 0.35 were prepared in air to stabilize the orthorhombic phase and suppress the occurrence of the hexagonal phase preferred for Y rich compositions [69]. Samples with $x = 0.4, 0.45, 0.5$ and 0.6 prepared by solid state reaction techniques always showed either a fraction of hexagonal phase or a high amount of Mn^{4+}. To avoid these phases the samples with $x \geq 0.4$ were treated in a final step in a floating zone image furnace by re-crystallizing stoichiometric ceramic rods with diameter of approximately 8 mm under a pressure of 8 bar of argon as described in Section 3.2. Crystalline boules from these experiments were crushed and ground to fine powder. A short overview of the samples and their preparation method is shown in Table 3.1.

3.2 Single Crystal Growth of $Nd_{1-x}Y_xMnO_3$

Single crystals of $Nd_{1-x}Y_xMnO_3$ were grown using the floating zone method in a two mirror image furnace, according to methods known from different rare earth manganites[77, 78,

---
[1]Higher temperature stabilizes the trivalent state of Mn^{3+}, Mn^{4+} appears more often at lower temperature inducing Mn vacancies in the crystal

3.2. Single Crystal Growth of $Nd_{1-x}Y_xMnO_3$

x	preparation method
0.0	SSR in N_2
0.1	SSR in N_2
0.2	SSR in N_2
0.3	SSR in air
0.35	SSR in air
0.4	FZ in Ar
0.45	FZ in Ar
0.5	FZ in Ar
0.6	FZ in Ar

Table 3.1: *List of the samples with the respective preparation conditions. Samples were prepared by solid-state reactions (SSR) in either N_2 or in air and by the optical floating zone (FZ) method in Ar atmosphere.*

79, 80, 81] (instrument description in Section A.14). Polycrystalline rods for crystal growth were prepared as described in Section 3.1. The rods (approximately 10 g with a diameter of 8 mm) were pressed at 300 bar in an isostatic press after the second firing followed by a firing at 1450°C, always using air as atmosphere.

3.2.1 The Floating Zone Method

The basic idea in floating zone crystal growth is to move a liquid zone through a polycrystalline material with the chance that the liquid solidifies in a single crystal phase. The first floating zone crystal growths were done for the production of high-quality Si single crystals. If properly seeded and under optimized conditions, a single crystal may result on the seed rod, as it is illustrated in Figure 3.1 and displayed in Figure 3.2.

The process of crystal growth works as shown schematically in Figure 3.3. The solid but polycrystalline feed rod is moved slowly in the hot zone of the focus of the infrared lamps. The power of the lamps is chosen so that the temperature at this spot is right above the melting point of the feed rod. As a consequence the volume of the liquid is increased. To keep the balance the seed rod is moved slowly out of the hot zone, ideally with the same speed as the feed rod. This moves also the bottom part of the liquid slowly out of the hot zone. The liquid begins to solidify at the surface of the seed rod and, if the movement is slow enough, the atoms in the liquid organize in a regular manner according to the crystalline surface of the seed rod leading at the end to a single crystal.

The method was first used for purification (zone melting), taking advantage of the small segregation coefficients k_{seg} of many impurities. The k_{seg} in the thermodynamic

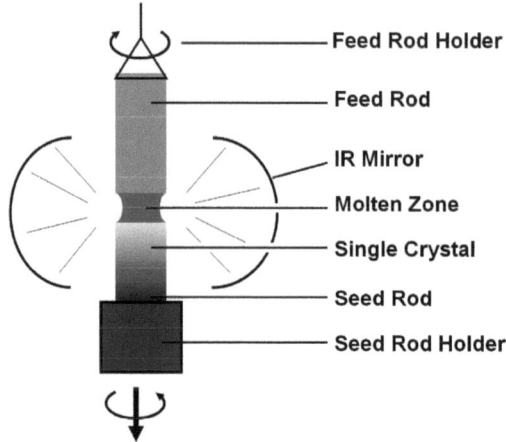

Figure 3.1: Schematic display of the floating zone process.

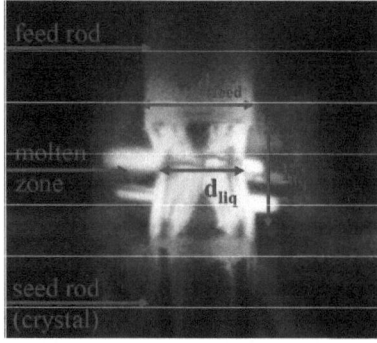

Figure 3.2: Picture taken from an actual growth. The molten zone is located where the infrared beams converge on one hot spot. The filament of the lamps can be seen in the reflection, visible as two white horizontal lines behind the melt. Ideally the the ratio of d_{liq} to d_{feed} should be about 0.8 and l_{liq} to d_{feed} about 1.

equilibrium[2] is defined by the ratio of the concentration of impurity atoms in the growing crystal and that of the melt. It is usually much lower than 1 because impurity atoms

---
[2]'Equilibrium' refers to a growth speed of 0 $\frac{mm}{h}$ or, more practically, very low growth rates. For finite growth rates, k_{seg} becomes a function of the growth rate (called k_{seff}) and approximates 1 for high growth rates (whatever comes to the rapidly moving interface gets incorporated).

3.2. Single Crystal Growth of $Nd_{1-x}Y_xMnO_3$

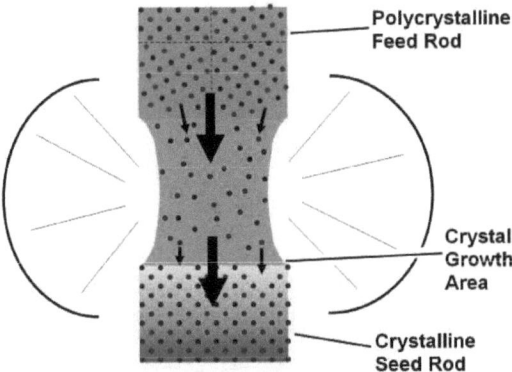

Figure 3.3: *This image displays schematically the polychrystalline feed rod with only local order, the liquid zone in the focus of the infrared lamps and the crystal grown on the feed rod.*

'prefer' to stay in the melt. For the well known process of silicon crystals example values are $k_{seg} = 10^{-6}$ for Metals like Ag, Au, Mn or Fe but only $k_{seg} = 1.25$ for oxygen, demonstrating the importance of the growth atmosphere.

So the impurities contained in the feed material would then prefer to remain in the melt and thus could be swept to the end of the feed stock. On one hand the crystal will be cleaner than the liquid, therefore crystal growth is simultaneously a purification method[3]. The concentration of impurities in the melt changes with the proceeding of the growth resulting in a change of the impurity content along the length of a crystal. It is obvious that a clean and well prepared feed rod is very important for homogeneous single crystals using this method.

If properly done, the newly crystallizing material can be obtained as a single crystal. The melt has contact only to vacuum or purified atmospheres, so there is no incorporation of impurities that the melt can pick up, in contrast to the solid state reaction technique used in Section 3.1 where the reacting components may interact with the crucible material and the flux methods where the flux is incorporated into the sample.

The problem of floating zone crystal growth is to keep the liquid between the feed rod and the seed rod. The main force to keep the liquid at place is surface tension. Usual solid state materials as transition metal oxides and rare earth oxides are limited to a maximum diameter of about 20 mm. Industrial facilities that grow silicium single crystals for microchips use further stabilizing mechanisms like drawing the liquid zone through a

---
[3]The last part of the crystal must be discarded as here all the impurities are now concentrated.

'hole'. However, for very large diameter crystals the difficulties grow rapidly and floating zone crystal growth is rarely used for diameters larger than 150 mm.

The crystal diameter is determined by the the temperature profile and the pressure applied. The hotter the melting zone, the larger the molten area and the less viscous is the liquid, resulting in the need for a larger area of the seed rod as shown in Figure 3.4. On the other hand high pressure increases the melting point to some degree with the need to increase the energy input to keep the state of the molten zone stable.

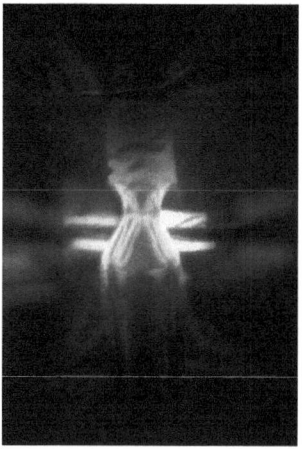

Figure 3.4: *Picture taken from a growth of NdMnO$_3$ with a molten zone too hot. The increasing diameter of the crystal with proceeding growth is visible.*

The condition of the preparation of the rod and the quality if the used atmosphere determines the quality and homogeneity of the crystal. It is very important to keep the conditions as stable as possible while growing because every disturbance can influence the quality of the crystal. The reason is that with starting the growth the liquid crystallizes at many points causing many grains to grow. The orientation of those is statistically distributed and the first part of a grown crystal is always a polycrystalline. Proceeding the growth usually one of the grains dominates and displaces the others with time and length resulting in the latter part of the growth to consist of only one grain.

An example of a successful growth is shown in Figure 3.5. Here the very sensitive steps of the seeding and the start of the growth is illustrated on the sample of $x = 0.5$.

3.2. Single Crystal Growth of $Nd_{1-x}Y_xMnO_3$

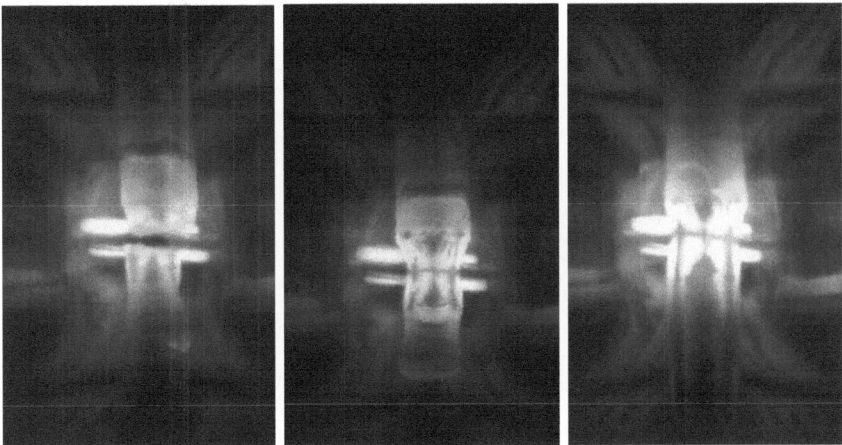

Figure 3.5: *Series of pictures taken from a growth of $Nd_{0.5}Y_{0.5}MnO_3$. Left the feed rod (top) and seed rod (bottom) are visible, rotating in opposite direction to each other, in the process of being heated up. In the middle feed and seed rod are connected with a molten zone in between. On the right the progress of the growth is visible with the crystal growing feed rod and seed rod moving downward with 2 mm per hour.*

3.2.2 Optimizing Conditions for Crystal Growth of $Nd_{1-x}Y_xMnO_3$

The conditions of a crystal growth in a mirror furnace determine the quality of the resulting crystal. The slower a crystal is grown, the more time the crystal has to crystallize in the right composition and crystal structure. This results in a crystal with less impurities and grains. As shown in Figure 3.1 the speed of the crystallizing process is determined by the time it takes to pull the feed and seed rod through the molten zone. A rod with the length of 10 cm and a growing speed of 2 $\frac{mm}{h}$ will take approximately 50 hours to grow. So it is obvious that time is a crucial factor and one objective is to grow the crystal as fast as possible and as slow as necessary.

A growth speed of 2 $\frac{mm}{h}$ was optimal as increasing the growth speed to higher rates resulted in highly polycrystalline samples, although it did not affect the hexagonal impurity ratio. As it can be seen in Figures 3.6 the change of the growth speed from 4 $\frac{mm}{h}$ to 2 $\frac{mm}{h}$ had a strong effect on the crystal outcome.

Using oxidizing or reducing atmospheres influences the valence state of the metals and therefore also the oxygen content. The pressure of the atmosphere though influences the partial pressure of oxidizing or reducing gases and has an impact on the reaction equilibrium. On the other hand the atmosphere pressure changes the melting temperature

Figure 3.6: *Result of a $x = 0.3$ growth at 1 bar of argon. The rod in the top image was grown with 4 $\frac{mm}{h}$, resulting in a highly polycrystalline outcome while the rod in the bottom image was grown with 2 $\frac{mm}{h}$, resulting in a crystal, though still with more than one grain.*

to some degree and can change the outcome by a higher or lower crystallizing temperature.

The polycrystalline precursor rods were made in air and had therefore always a potion of oxidized Mn^{4+}. Argon has a weak reducing property and was able to reduce the Mn^{4+} to Mn^{3+} when used as growing atmosphere. Growing at pressures of at least 8 bar supported the formation of the orthorhombic crystal structure, while crystals grown at 1 bar tended to have hexagonal impurities, whose amount increased with x.

Samples grown in floating zone that were used for measurement of physical properties and neutron diffraction are listed in Table 3.2, including the information of the neutron scattering instrument they were examined with.

3.2.3 Characterization of the Quality of Single Crystals

A perfect single crystal has a very high regularity with the right crystal structure, right stoichiometry, all atoms in the right valence state and with only one grain. The quality of the grown crystals was determined by eye, with x-ray powder diffraction and with Laue diffraction.

The appearance by eye of a grown crystal already gives some information of the quality.

3.2. Single Crystal Growth of $Nd_{1-x}Y_xMnO_3$

x	D10 2.36 Å	V2 5.0 Å	E4 2.47 Å 40' col.	E4 2.44 Å 20' col.	E5 2.4 Å	D10 Integrated Intensity
0.30	x					
0.35	x					
0.40		x	x		x	
0.45	x	x	x	x		x
0.50	x					
0.55	x		x	x		x

Table 3.2: List of the single crystal samples grown in a floating zone furnace and measured with different instruments at Helmholtz-Zentrum Berlin für Materialien und Energie and Institut Laue-Langevin. Details of the measurements are described in Section 2 and A.

Usually a shiny regular surface without visible grains counts for a good crystal while, on the other extreme, a matt locking rod counts for a polycrystalline crystal as seen in Figure 3.6 (top).

Performing x-ray powder diffraction at room temperature gives an insight in the different crystal structures in the material, at best only one compound can explain the whole diffraction pattern while impurities may serve further information, for example when hexagonal phases are detected. The diffraction pattern can be analyzed and crystallographic parameters like the cell size and the position and ratio of the heavy metals can be calculated. Although if the diffraction pattern shows only one compound that confirms the one for orthorhombic $Nd_{1-x}Y_xMnO_3$, the material must not be single crystalline and Laue diffraction patterns (for details see Section 2.1.1) must be done.

A single crystal Laue diffraction pattern shows sharp spots that can be explained with the reflections of one space group. Multiple grains show additional spots that are shifted by some degree. In high grainy material the spots get very diffuse until no pattern can be detected at all. Neutron Laue diffraction shows the pattern of a crystal volume as neutrons penetrate the crystal while x-ray Laue diffraction probes only the surface. In the latter case is is necessary though to make Laue patterns of many different positions of the crystal. If all the collected patterns can be explained with the same spacial orientation of the crystal it may be assumed that also the crystal volume is oriented in this direction and consists therefore of a single grain.

If the grown crystal is so confirmed with the right stoichiometry and as a single grain crystal this piece can be used for further treatment like for cutting, polishing and gold sputtering to prepare thin plates (as shown in Figure 3.7) for physical properties measurements.

The quality of the samples varied from very good (i.e. single grain crystal) to fair (i.e. crystal with multiple grains that are oriented in the same direction with a deviation of

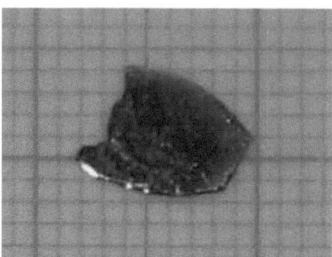

Figure 3.7: *Gold sputtered plate of $Nd_{0.55}Y_{0.45}MnO_3$ cut in the ab plane with a thickness of 0.9 mm.*

maximum 4 degree) as it is visible in Figure 3.8 from the neutron Laue images taken at OrientExpress in ILL.

3.2.4 Thesis objective

Although the suppression of the collinear A-type antiferromagnetic ordering in favor of cycloidal order for R = Tb and Dy is evident, the exact details of how one magnetic structure develops into the other are not entirely clear. For example is the transition from A-type to cycloidal ordering with decreasing R-size first or second order? Ideally one would perform neutron diffraction experiments to probe the magnetic ordering directly, however the high neutron absorption cross section of Sm, Eu and Gd, that make up compositions in the the cross-over region from A-type to incommensurate magnetic order, prohibit such experiments. While the solid solution of $Eu_{1-x}Y_xMnO_3$ has been examined, detailed neutron scattering experiments are prohibitive for low Y doping.

To examine the onset of the magnetic frustration in the RMnO$_3$ system $Nd_{1-x}Y_xMnO_3$ samples were synthesized as described in Section 3. Many properties count for this system to be chosen for examining the development of the magnetic properties with changing tolerance factor with neutrons.

First the fact that, as well as in the $Eu_{1-x}Y_xMnO_3$ system, the tolerance factor and therefore the <Mn-O-Mn> angle can be tuned to values different than the one elemental compounds. Second the neutron scattering cross section and the absorption cross section of Nd and Y ($\sigma_{scatt}(Nd) = 16.6$ barn, $\sigma_{abs}(Nd) = 50.5$ barn, $\sigma_{scatt}(Y) = 7.7$ barn, $\sigma_{abs}(Y) = 1.28$ barn) allow good neutron scattering results compared to Eu and Gd compounds with very high absorption cross sections ($\sigma_{abs}(Eu) = 4530$ barn, $\sigma_{abs}(Gd) = 49700$ barn). Third the magnetic moment of Nd^{3+}[4] is not very high and Y^{3+} has

---

[4]The Nd^{3+} ion consists of $3f$ electrons. Their orbital and spin movements are highly correlated and for the Nd^{3+} ion the lowest multiplet given by Hund's rules is $^4I_{\frac{9}{2}}$ with J = 9/2, S = 3/2, L = 6 and the

3.2. Single Crystal Growth of $Nd_{1-x}Y_xMnO_3$

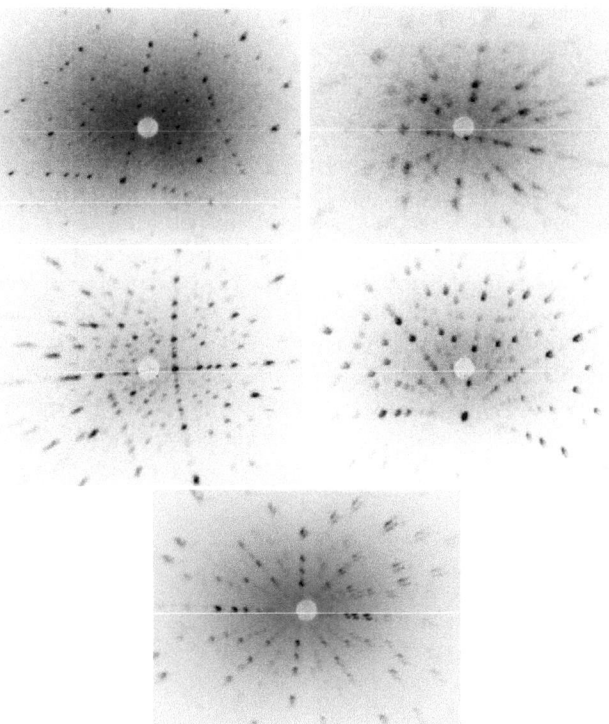

Figure 3.8: *Neutron Laue images of samples grown in floating zone procedure. From top left (tl) to bottom (b) the samples are $x = 0.3$ (tl), 0.35 (tr), 0.4 (ml), 0.45 (mr) and 0.5 (b).*

none and though does not influence the magnetic order of the Mn^{3+} as much as other rare earth ions like in Eu^{3+} or Tb^{3+}. For example in $TbMnO_3$ Tb orders below $T_N^{Tb} = 6.5$ K resulting in a decreased ferroelectric polarization in the ferroelectric regime, although this is a very small effect.

To examine the magnetic transitions of $Nd_{1-x}Y_xMnO_3$ with changing x polycrystalline samples and single crystalline samples were synthesized and analyzed as described in Section 3 and 2. The polycrystalline samples were easier to make as it only needed solid state chemistry but there are a few drawbacks. Powder samples are by nature macroscopically isotropic and a measured property like magnetism or polarization is always the mean value of the anisotropic properties in the three spatial directions. Powder diffraction allows to

---

Landé factor g = 8/11.

get a good insight in the crystal structure but the expected magnetic peaks may overlap with the nuclear peaks, making it difficult to distinguish the peaks and to analyze the position and the peak shape.

Single crystals on the other hand can be aligned and cut, the anisotropic physical properties are measurable in any direction and all reflection selective neutron scattering experiments are applicable. On the other hand every reflection has to be measured separately. To measure a broad set of different reflection takes many measurements and therefore much longer compared to a powder diffraction where a one diffraction pattern covers a range of 2θ from 0 to 120 degree in a fast single measurement.

To map out the general development of the crystal structure, the general magnetic properties and the magnetic structure in the $Nd_{1-x}Y_xMnO_3$ system first polycrystalline samples were synthesized and analyzed and will be the topic in Section 4 as it was published by Landsgesell et al.[82]. Single crystal were grown, oriented and partly cut for analyzing the magnetic, dielectric and pyroelectric properties in the PPMS as well as determining the detailed magnetic structure using multi axial neutron diffraction and will be discussed in the Sections 7, 5 and 6.

Chapter 4
Characterization of Crystal and Magnetic Structure of $Nd_{1-x}Y_xMnO_3$

By changing the rare earth element in $R MnO_3$ the magnetic structure changes from collinear A-type antiferromagnet to E-type antiferromagnet crossing an area of frustrated magnetism. This is responsible for a phase transition from a collinear spin density wave to a non collinear spin cycloid causing a ferroelectric polarization. The development of the spin cycloid from collinear A-type antiferromagnet and the induced ferroelectric polarization is studied by varying x in $Nd_{1-x}Y_xMnO_3$.

Using polycrystalline samples of $Nd_{1-x}Y_xMnO_3$ it is possible to get an overview of the the structural and magnetic transition with changing x. Powder diffraction data can provide crystallographic reflections over a wide range of 2Θ with only one measurement as well as information about impurities and, depending on x-rays or neutron diffraction, the stoichiometry of the atoms. With temperature dependent magnetic susceptibility and neutron diffraction measurements also the transition temperatures can be determined. With the information gained with polycrystalline samples single crystal growth can be optimized. Further, after the general crystal and magnetic structure is known, the single crystal neutron diffraction measurements can be focused on the important reflections.

4.1 Isotropic Magnetic Properites

Data were measured on warming from 2 K to 200 K for $x = 0.0$ to 0.6 polycrystalline samples using the SQUID magnetometer (see Section A.1) and the D20 neutron diffractometer (see Section A.13)

The magnetization data shown in Figure 4.1 for the $x = 0$ sample show a sharp up-turn at 87 K, a behavior that is not expected for antiferromagnetic ordering, neither reminiscent of a second order phase transition, although the onset of this feature is consistent with value of T_N from previous reports[83, 84]. The magnetization data is reminiscent of an induced ferromagnetic transition likely arising from the ordering of Nd^{3+} spins[83, 84]. As discussed elsewhere[83, 85], the A-Type order of the Mn^{3+} spins exhibits a non negligible antisymmetric component leading to a small ferromagnetic component perpendicular to

x	remanence $\left(\frac{emu}{mol}\right)$	moment $\left(\frac{\mu_B}{f.u.}\right)$	T_N ($\chi(T)$) (K)	T_N (NPD) (K)
0.0	7.7	1.38	88	-
0.1	5.7	1.02	83	-
0.2	4.8	0.86	75	-
0.3	3.7	0.66	63	-
0.35	2.8	0.50	56	57
0.4	2.1	0.38	54	55
0.45	1.8	0.32	53	52
0.5	0.6	0.10	52	51
0.6	-	-	50	49

Table 4.1: List of the magnetic remanence and effective moment of the polycrystalline samples. Ferromagnetic remanence and moments were determined by isothermal magnetic hysteresis measurements at 5 K, while the effective moment was determined using temperature dependent magnetization. Néel temperatures (T_N) determined from magnetization ($\chi(T)$) and temperature dependent neutron powder data (NPD) measurements are also given.

the sheets (i.e. $H \| c$). Therefore the A-type ordering of Mn-spins can likely induce the ferromagnetic ordering of Nd^{3+} spins. As a result T_N and T_C are at the same temperature and the rise of the magnetization curve should be indicative of T_N. This assumption is confirmed by our neutron diffraction measurements described below, where Rietveld refinements of the data indicate that the Mn spins order with A-type structure and Nd-spins order ferromagnetically along the c-axis.

The ferromagnetic ordering of Nd-spins is also reflected in hysteresis curves measured at 5 K as displayed in Figure 4.2. The $x = 0$ sample show a magnetic behavior typical for a weak ferromagnet with a coercivity of lower than 1000 Oe and a small but measurable remanence, while hysteresis loops measured at 85 K were consistent with a paramagnetic behavior. With increasing x, the strength of the coercive field is approximately the same for all samples but the remanence field and saturation (ferromagnetic) moment decreases with x as indicated in Table 4.1.

The sample $x = 0.6$ does not show a ferromagnetic hysteresis loop, indicating that for this composition there is no significant ferromagnetic contribution to the magnetic ordering. The susceptibility measurements from all samples followed a Curie-Wiess behavior above T_N giving Curie constants that decrease with increasing x and vary from from $\mu_{eff} = 5.8$ to 5.4 μ_B, somewhat lower than the expected values of $\mu_{eff} = 6.1$ to 5.7 μ_B.

The magnetization data show that with decreasing x the onset temperature of the magnetic ordering also decreases, from $T_N = 88$ K to 49 K as x is varied from 0 to 0.6. Here the values of T_N or T_C were determined by taking the derivative of $M(T)$ and are given in Table 4.1. The overall behavior of the $Nd_{1-x}Y_xMnO_3$ solid solution therefore

4.1. Isotropic Magnetic Properites

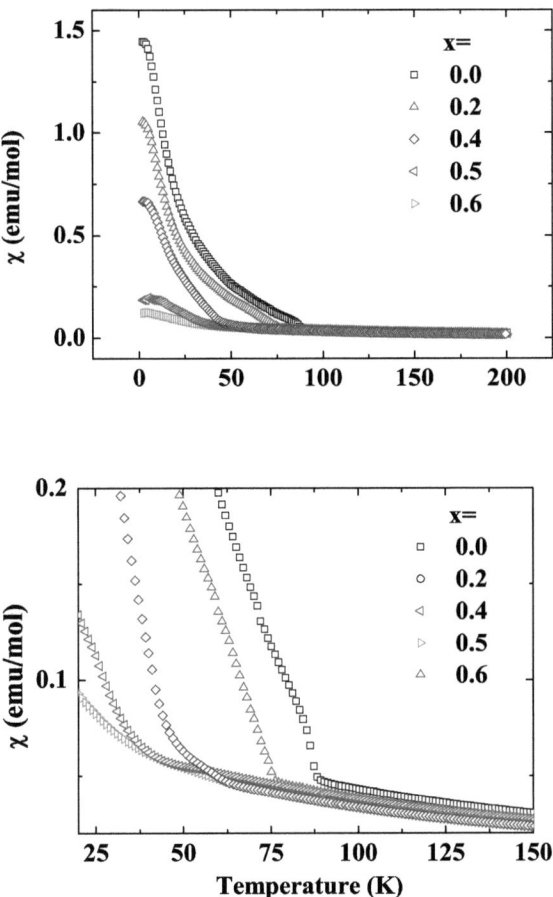

Figure 4.1: *Temperature dependent dc magnetization measurements from samples $x = 0.0$, 0.2, 0.4, 0.5 and 0.6, performed using a field of 5000 Oe. The dc magnetization measurements for the remaining samples are not shown for clarity, but follow the same trends as indicated by the data shown.*

tracks that of the $R\text{MnO}_3$ perovskite manganites where T_N is reduced as the size of the rare-earth ion is decreased[40].

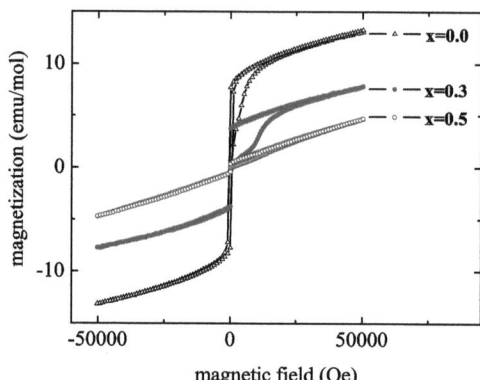

Figure 4.2: *Hysteresis curves measured with dc SQUID magnetometer at 5 K on samples $x = 0.0, 0.3$ and 0.5.*

4.2 Room Temperature Crystal Structure

To show that by varying x the crystallographic and macroscopic magnetic properties change comparable to RMnO$_3$ the results of the magnetic measurement and the Rietveld refinements of the polycrystalline samples at room temperature are presented here. The measurements were done on E9 and D20, for instrumental details see Section A.6 and A.13. The analysis of the x-ray and neutron powder diffraction patterns was done with the Rietveld method using the program WinPLOTR Fullprof suite

Both x-ray- and neutron powder diffraction data measured at 300 K of the samples described in Section 3.1 were consistent with an orthorhombically distorted perovskite structure with space group P*bnm* (No. 62)[86]. The extinction rules for the space group P*bnm* are listed in Table 4.2.

Fig. 4.3 shows a representative Rietveld fit to the neutron powder diffraction data measured at 300 K from the $x = 0.45$ sample, while Table 4.3 reports relevant crystallographic parameters at 300 K for all samples.

As it is evident from Tables 4.3 and 4.4, changes of lattice constants and atomic positions are consistent with Vegard's law[1] indicating that the substitution of Y for Nd results in a

---

[1]Vegard's law is an approximate empirical rule which holds that a linear relation exists, at constant temperature, between the crystal lattice constant of a composition, with atomic positions being partially replaced in a solid solution, and the concentrations of the constituent elements. Applications demonstrate the importance of relative atomic sizes in determining lattice constants and suggests that for sufficiently

4.2. Room Temperature Crystal Structure

Multiplicity and Wyckoff letter	Reflection conditions
8d	0kl : k + l = 2n
	h0l : h = 2n
	h00 : h = 2n
	0k0 : k = 2n
	00l : l = 2n
4c	no extra conditions
4b	hkl : h + k, l = 2n
4a	hkl : h +k, l = 2n

Table 4.2: *Extinction rules for the Pbnm system at different crystallographic positions.*

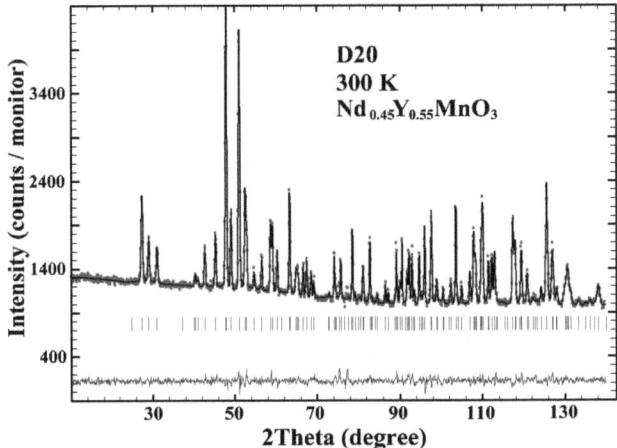

Figure 4.3: *A typical Rietveld analysis of the neutron powder diffraction data. The neutron powder diffraction data (circles) were measured from the $x = 0.45$ sample at 300 K using the D20 diffractometer (see Chapter A.13). The upper solid line represents the calculated neutron powder diffraction pattern and the lower solid line the difference between measured and calculated patterns. The vertical tick marks represents allowed reflections for the Pbnm space group.*

solid solution. Careful examination of peak profiles of the the neutron powder diffraction data gave no indication of the presence of two orthorhombic phases indicative of chemical phase separation or a miscibility gap. The values of the full width half maximum from

---

small disparities in atomic size Vegard's law may also hold along the fluid-solid coexistence curve.

4. Characterization of Crystal and Magnetic Structure of $Nd_{1-x}Y_xMnO_3$

x	a (Å)	b (Å)	c (Å)	χ^2
\multicolumn{5}{c}{polycrystalline (N_2)}				
0	5.4146 (3)	5.8472 (3)	7.5445 (4)	1.90
0.1	5.4006 (3)	5.8544 (3)	7.5280 (4)	2.02
0.2	5.3955 (2)	5.8533 (2)	7.5120 (2)	2.30
\multicolumn{5}{c}{polycrystalline (air)}				
0.3	5.3587 (4)	5.8226 (4)	7.4841 (4)	2.86
0.35	5.3540 (3)	5.8319 (3)	7.4771 (4)	2.24
\multicolumn{5}{c}{crystalline samples}				
0.4	5.3549 (3)	5.8624 (3)	7.4666 (4)	3.39
0.45	5.3428 (3)	5.8617 (3)	7.4530 (4)	2.24
0.5	5.3436 (3)	5.8639 (3)	7.4542 (4)	6.8
0.6	5.3275 (2)	5.8607 (2)	7.4370 (3)	8.7

Table 4.3: Room temperature structural parameters of the unit cell in Pbnm symmetry and R-factors determined from Rietveld analysis of the neutron powder diffraction data.

x	x (Nd/Y)	y (Nd/Y)	x (O1)	y (O1)	x (O2)	y (O2)	z (O2)
\multicolumn{8}{c}{polycrystalline (N_2)}							
0	-0.0141 (7)	0.0708 (6)	0.0901 (7)	0.4735 (7)	0.7147 (6)	0.3167 (6)	0.0478 (6)
0.1	-0.0142 (9)	0.0714 (8)	0.0917 (9)	0.4721 (8)	0.7144 (8)	0.3189 (8)	0.0476 (8)
0.2	-0.0143 (7)	0.0736 (7)	0.0938 (7)	0.4720 (7)	0.7127 (6)	0.3204 (6)	0.0477 (6)
\multicolumn{8}{c}{polycrystalline (air)}							
0.3	-0.0137 (7)	0.0757 (7)	0.0911 (7)	0.4735 (8)	0.7065 (7)	0.3228 (6)	0.0481 (6)
0.35	-0.0139 (8)	0.0768 (8)	0.0924 (8)	0.4744 (7)	0.7056 (7)	0.3230 (7)	0.0483 (7)
\multicolumn{8}{c}{crystalline samples}							
0.4	-0.0152 (7)	0.0768 (7)	0.0958 (7)	0.4711 (6)	0.7087 (6)	0.3233 (6)	0.0488 (6)
0.45	-0.0154 (6)	0.0778 (6)	0.0976 (6)	0.4703 (6)	0.7074 (5)	0.3244 (5)	0.0490 (5)
0.5	-0.0157 (9)	0.0788 (9)	0.0991 (9)	0.4708 (8)	0.7065 (8)	0.3246 (8)	0.0491 (8)
0.6	-0.0166 (9)	0.0806 (9)	0.1015 (9)	0.4688 (8)	0.7046 (8)	0.3258 (8)	0.0498 (8)

Table 4.4: Room temperature structural parameters of the atomic sites. The general positions, according to Table 4.2, are: Mn^{3+} at $4b(\frac{1}{2}00)$, Nd^{3+}/Y^{3+} at $4c(xy\frac{3}{4})$ and the O^{2-} atoms (O1 and O2) at $4c(xy\frac{1}{4})$ and $8d(xyz)$.

the neutron powder diffraction data were close to those expected from the instrumental resolution function and making it possible to detect possible chemical phase separation at $\Delta x \geq 0.05$.

While diffraction data from all samples could be modeled using a single phase orthorhombic perovskite phase, measurements of the $x = 0.5$ and 0.6 samples showed small traces of the hexagonal manganite phase (0.9 and 1.4 weight %, respectively). This observation indicates that the orthorhombic to hexagonal phase boundary is located approximately around $x \approx 0.55$ and varies with preparation conditions. Indeed it was not possible to synthesize phase pure samples with $x > 0.6$, and the best results in this high x region were for samples prepared by the floating zone method. This behavior is similar to that found in the $Eu_{1-x}Y_xMnO_3$ solid solution where this phase boundary occurs at $x \approx 0.7$[67]. In general all three lattice constants decrease with increasing x reflecting the smaller size of the Y^{3+} ion substituted for the larger Nd^{3+}.

The difference of the ionic size between Nd^{3+} and Y^{3+} is significant (98 pm and 90 pm, respectively) with the ionic variance reaching a maximum at $x = 0.5$ and the unit cell volume and lattice constants decreasing with tolerance factor and mean ionic radii, $\langle r_A \rangle$ of the R-site. If composition x is scaled to $\langle r_A \rangle$, then the $x = 0.4$ sample has the same $\langle r_A \rangle$ as $EuMnO_3$. However, the volume and the lattice constants match those of $x = 0.5$. Also the same behavior is found for $SmMnO_3$ where the $x = 0.25$ sample has the same $\langle r_A \rangle$, but its volume matches that for $x = 0.35$. This is indicative for a systematic departure from the trends made up for the $RMnO_3$ series, as was noted also for $Eu_{1-x}Y_xMnO_3$[67].

Fig. 4.4 displays the variation of the <Mn-O-Mn> bond angles and <Mn-O> bond lengths as a function of x determined from the neutron powder diffraction data. The <Mn-O> bond lengths show an almost negligible dependence on x, a result that is not surprising considering that they are sensitive to the Mn^{3+} oxidation state which remains unchanged for this series of compounds. Indeed the 300 K neutron powder diffraction data is consistent with the orbital ordering of Mn^{3+} $3d_{3x^2-r^2}$ and $3d_{3y^2-r^2}$ orbital as found in $LaMnO_3$[87]. On the other hand the <Mn-O-Mn> bond angle depends on x, it decreases linearly from $149.0(1)°$ to $146.1(1)°$ as x is varied from 0 to 0.6 and reflects the decrease of $\langle r_A \rangle$. Indeed in the $RMnO_3$ manganites a similar trend is found with decreasing r_A. As R is varied from Nd through to Tb, the <Mn-O-Mn> bond angle varies from $149°$ to $143°$, and r_A varies over a similar range as it is an $Nd_{1-x}Y_xMnO_3$ solid solution (i.e. from 98 pm for Nd^{3+} to 92 pm for Tb^{3+})[40]. This suggests that the effects of spin frustration due to the structural tuning of the J_1 and J_2 interactions in the $Nd_{1-x}Y_xMnO_3$ solid solution should be manifested in a similar way to the $RMnO_3$ series of compounds.

4.3 Magnetic Structure

Temperature dependent neutron powder diffraction measurements were performed to characterize the general magnetic ordering of the $Nd_{1-x}Y_xMnO_3$ solid solution samples. Data were measured on warming from 2 K to ≈ 120 K for $x = 0.35$ to 0.6 samples using the D20 diffractometer (see Section A.13), while further data were measured at 2 K for $x = 0.0$ to 0.3 samples on the E9 diffractometer (see Section A.6). The values for T_N of the magnetization measurements can be confirmed using the temperature dependence of magnetic

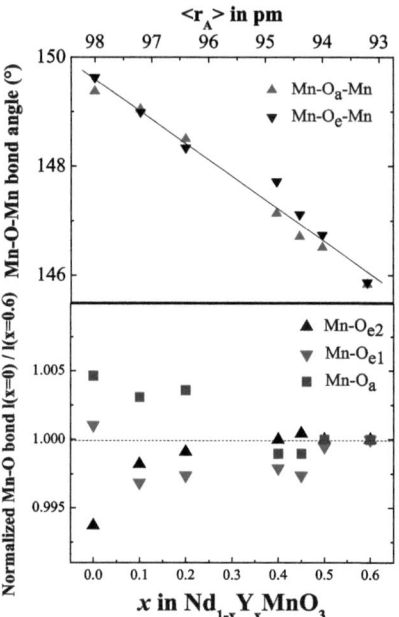

Figure 4.4: Variation of the <Mn-O-Mn> angles (upper frame) and the three <Mn-O> distances (lower frame) with x (lower x-axis) and mean ionic radii $\langle r_A \rangle$ in the solid solution $Nd_{1-x}Y_xMnO_3$. The <Mn-O> bond lengths are normalized to those determined for $x = 0.6$. In this orthorhombic perovskite there are three <Mn-O> bond lengths and two <Mn-O-Mn> angles. The subscript 'e' and 'a' denotes equatorial and apical bond lengths and angles. Error bars are smaller or equal in magnitude to the size of the symbols.

intensities. Both onset values agree well as illustrated by the example scan for the $x = 0.4$ sample shown in Figure 4.5 and Table 3.1.

More importantly for the $x \leq 0.3$ samples, the quality of the Rietveld analysis could be improved significantly by the addition of a ferromagnetic ordering for the Nd^{3+} ions for samples with $x = 0$ to 0.3 as shown in Fig. 4.6, confirming the conclusions of the magnetization measurements. The largest ferromagnetic contribution is observed for the $x = 0$ sample where Nd^{3+} spins order along the c-axis with a moment of 1.5 μ_B/Nd. As x

4.3. Magnetic Structure

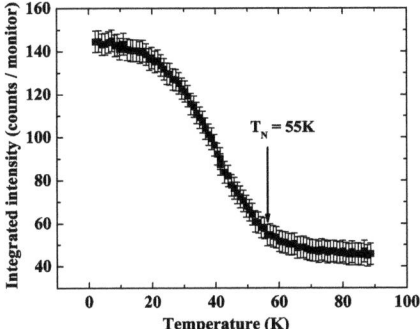

Figure 4.5: *Temperature dependence of the integrated intensity of the (001) magnetic reflection indicative of A-type antiferromagnetic ordering. These neutron powder diffraction data were measured from the $x = 0.4$ sample using the D20 diffractometer. T_N as indicated by these data is shown and agrees well with the value of 54 K determined from the magnetization data. For the neutron powder diffraction data T_N was taken as the temperature at which magnetic Bragg peaks disappeared on warming.*

increases the magnitude of the Nd^{3+} moment steadily decreases as displayed in Figure 4.9, while for $x > 0.3$ no evidence of ferromagnetic Nd^{3+} ordering in the neutron powder diffraction data can be found. The Rietveld analysis of the neutron powder diffraction data for $x \leq 0.2$ compositions, indicate that Mn-spins are collinear and oriented along the b−axis (R_w= 6.0% for moment parallel to b−axis ($m\|b$) compared to $R_w = 8.2\%$ for $m\|a$ or $R_w = 11.9\%$ for $m\|c$).

Higher x compositions show a series of incommensurate magnetic reflections below T_N in addition to the A-type magnetic Bragg peaks. These incommensurate reflections can be indexed with a magnetic propagation vector of $k = q\mathbf{b}^*$ with $q = \{0.23\ldots 0.27\}$, what is in the region of the known $TbMnO_3$ that exhibits at ground state a $q = 0.27$[88, 53]. This change in magnetic order with x is illustrated in Fig. 4.7 showing neutron powder diffraction data measured over the angular region $2\theta = 12\ldots 18$ at 2 K for compositions $x = 0.35$ to 0.6. This angular range covers the region where the commensurate (001) reflection indicative of A-type order and the incommensurate $(0\pm q1)$ reflections are observed. For the $x = 0.35$ sample a single (001) reflection is found, however for increasing x there is an additional reflection that appears at slightly higher angles and for $x = 0.6$ only this incommensurate reflection is observed. For the intermediate compositions, $0.4 \leq x \leq 0.5$ the intensity ratio of these two reflections varies, favoring the incommensurate peak as x

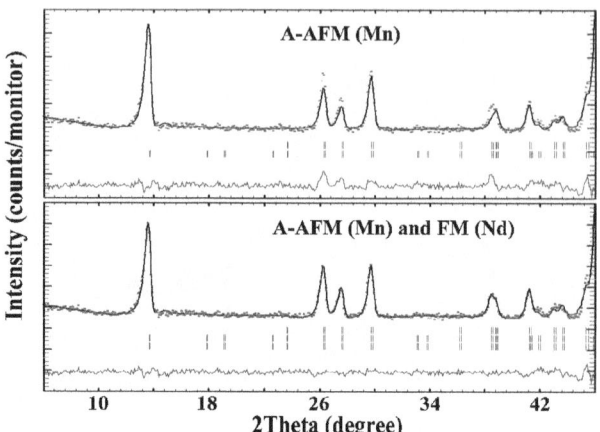

Figure 4.6: Rietveld analysis of the neutron powder diffraction data measured from NdMnO$_3$ at 2 K using the D20 diffractometer displayed over the region of $2\theta = 6°$ to $46°$. In the upper panel the data is modeled assuming an A-type ordering for Mn spins only. This model of nuclear and magnetic structure leads to a deficiency in the intensity of reflections between $26°$ and $29°$. In the lower panel results of an analysis using the same model as above with the addition of a ferromagnetic Nd-spin component along the c−axis. This model yields a magnetic moments of 4.1 and 1.5 μ_B per formula unit for the Mn and Nd ions respectively.

increases. A notable feature of these data is that for $x = 0.4$ the incommensurate reflection appears broader than the commensurate (001) peak, while in the $x = 0.5$ sample the reverse behavior may be occurring, and for the $x = 0.6$ sample the incommensurate reflection is as sharp as the (001) peak in the $x = 0.35$ sample. This may indicate that the coherence length of these two different magnetic orderings may be varying through this region of coexistence. Due to the high degree of peak overlap it is difficult to reliably quantify this observation with powder data and will be further examined in Section 7. Temperature dependent neutron powder diffraction data indicates that T$_N$ for both commensurate and incommensurate reflections is the same.

For the $x = 0.6$ sample, the neutron powder diffraction data were successfully modeled using a simple amplitude modulation of Mn-spins directed along the b−axis. The magnetic incommensurate reflections obeyed the extinction condition $h+k$=even, and l=odd, which correspond to an A-mode using the notation of Bertaut and Brinks et al.[89, 90]. The spin density wave model used to successfully analyze these data was of the irreducible repre-

4.3. Magnetic Structure

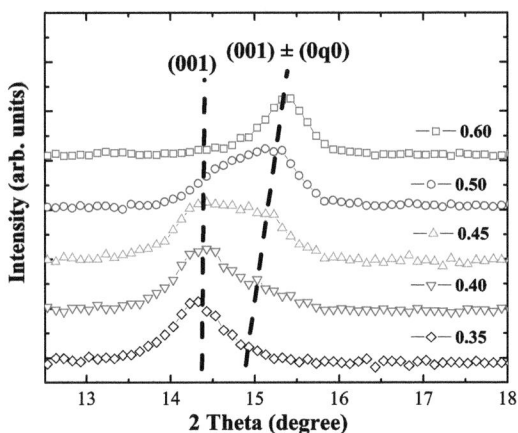

Figure 4.7: Portion of the neutron powder diffraction data measured on D20, showing the antiferromagnetic (001) A-type reflection and the incommensurate spin density wave reflection (001)±(0,q,0) for various samples measured at 2 K.

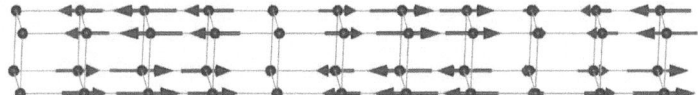

Figure 4.8: Modulation of an A-type antiferromagent with a commensurate spin density modulation of $q = 0.125$ along the b-direction

sentation[2] $\Gamma_3(0, A_y, 0)$ (see Fig. 4.8) and lead to a satisfactory refinements of the magnetic ordering below T_N, giving a saturated magnetic moment of ≈ 4.2 μ_B ($R_w(m\|b) = 22.8\%$), a value slightly higher than ≈ 4.0 μ_B expected for Mn^{3+}. Significantly lower quality fits to the data were obtained for models where Mn-spins are collinear and point along the $a-$ or $c-$direction yielding $R_w(m\|a) = 24.4\%$ or $R_w(m\|c) = 28.9\%$)[3], respectively. Attempts to fit a cycloidal model to these data ($\Gamma_2 \times \Gamma_3(0, A_y, A_z)$) were not statistically meaningful as powder data are incapable of uniquely differentiating between orthogonal spin components with the same extinction conditions [90]. This makes it necessary to use single crystal neutron diffraction to solve the magnetic structure as it is shown in Section 7.

---
[2] For details on the irreducible representations see Section 7.1.
[3] The higher R-factors in the $x = 0.6$ sample are due to the mentioned impurities in this sample

The neutron powder diffraction data was analyzed using these two models for the A-type and spin density wave phases in the region of co-existence with $0.4 \leq x \leq 0.5$. These two models described the neutron powder diffraction data very well and tests also confirmed the conclusion that in both phases the Mn-moments are aligned along the b-axis. Considerably worse fits to the data [4] resulted if the the $x = 0.45$ sample was tested with models where the incommensurate component was placed perpendicular to k. Measurements on single crystals will show that this assumption is correct as the magnetic peaks can be measured separately for they occupy different places on the Ewald sphere. As the commensurate and incommensurate magnetic phase can not occupy the same physical volume, because their spin components are parallel, this suggests that this intermediate region of composition between the purely A-type and spin density wave phases, is characterized by a magnetic phase separation.

The weight % variation of the two magnetic phases can be estimated across the region of co-existence by assuming that the ordered Mn moment in the region of $0.3 \leq x \leq 0.45$ is the same as for the $x = 0.2$ and 0.6 samples for the A-type and spin density wave phases respectively at 2 K. This makes it possible to fix the value of the ordered moment in the phase separated region and refine the magnetic phase fraction that can be represented as wt% (Fig. 4.9). For the samples $x = 0.0$, 0.1 and 0.2 there is no phase separation, although the ordered Mn moment that can be computed from the Rietveld analysis is decreasing slightly with increasing x, as does the Nd^{3+} ordered moment. These samples would represent (as deduced from the current neutron powder diffraction data) 100% of A-type phase. In the region of $0.3 \leq x \leq 0.45$ the relative amount of the A-type phase decreases while the spin density wave phase replaces it and for the $x = 0.6$ sample only the spin density wave phase is observed.

4.4 Phase Diagram of the Polycrystalline Samples

The magnetization and neutron powder diffraction data of the polycrystalline samples are presented in the magnetic phase diagram shown in Fig. 4.10. In general in the $Nd_{1-x}Y_xMnO_3$ solid solution, increasing x leads to a decrease of T_N for A-type ordering as well as T_C for Nd^{3+} ferromagnetic ordering. The reason for this decrease is different for the two magnetic ions. The strongest influence on the Nd^{3+} ordering is the dilution of this site with non-magnetic Y^{3+} ions. Although a ferromagnetic Nd component can not be detected in the neutron powder diffraction data for $x > 0.3$, the magnetization measurements suggest the presence of weak ferromagnetism for x as high as $x = 0.5$, while no ferromagnetism is detected for $x = 0.6$, within the sensitivity of our polycrystalline

---

[4] R_w=45 and 22% for A_x and A_z incommensurate components, compared to R_w=16% for the refinement of the A_y component

4.4. Phase Diagram of the Polycrystalline Samples

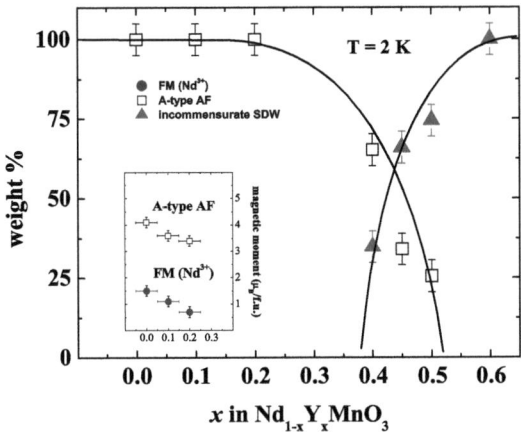

Figure 4.9: Relative amounts (in weight%) for the two magnetic phases found in the $Nd_{1-x}Y_xMnO_3$ solid solution at 2 K. The relative amounts of these magnetic phases is determined by assuming that in the mixed phase region the magnetic ordering is the same as in the $x = 0.2$ and 0.6 samples for the A-type and spin density wave phases respectively. While the amplitude of the Mn spin remains fixed in the analysis of the $x = 0.45$, 0.5 and 0.55 samples, the magnetic phase fraction is allowed to vary. For the remaining compositions where a single magnetic phase was observed, the magnetic scale factor was set to equal that of the structural nuclear scale factor. In the inset the results of the size of the Nd and Mn magnetic moments is shown determined from the Rietveld analysis of the neutron powder diffraction of composition with $x \leq 0.2$.

measurements.

The decrease of T_N for the A-type ordering is strongly coupled to the changes in the <Mn-O-Mn> angles that accompany the reduction of $\langle r_A \rangle$. Indeed this behavior tracks well the magnetic phase diagrams of $RMnO_3$ (see Figure 4.10) [40] and $Eu_{1-x}Y_xMnO_3$ [67] systems and suggests that the decrease of T_N in $Nd_{1-x}Y_xMnO_3$ is due to the (average) structural tuning of the relative strength and sign of nearest and next nearest magnetic interactions. The similarities between the phase diagrams of the $RMnO_3$ and $Nd_{1-x}Y_xMnO_3$ systems is not limited to the decrease in T_N. The observation of a spin density wave magnetic ground state for higher x compositions, coincides with $\langle r_A \rangle$ values similar to those for $TbMnO_3$ which exhibits spin density wave order below $T_N \approx 41$ K and cycloidal order below 28 K. Indeed the direction and magnitude of the incommensurate propagation

4. Characterization of Crystal and Magnetic Structure of $Nd_{1-x}Y_xMnO_3$

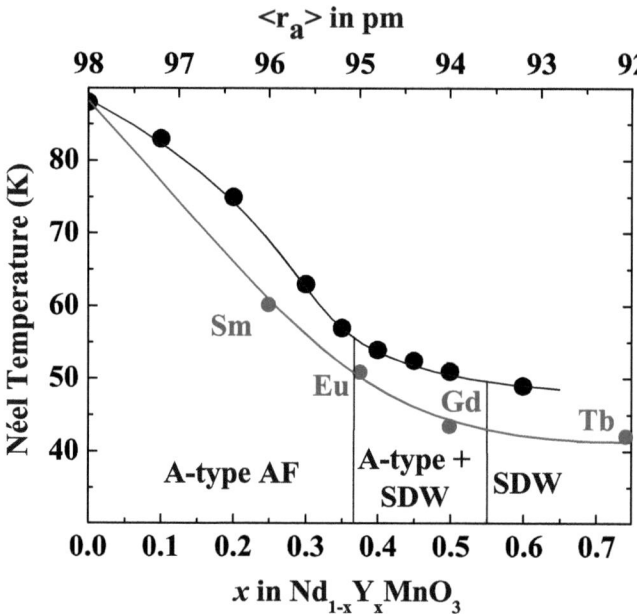

Figure 4.10: Magnetic phase diagram for the $Nd_{1-x}Y_xMnO_3$ solid solution determined from this work. The Néel temperature for the $RMnO_3$ series of manganites is also shown and is taken from reference [19].

vector is very similar in the $Nd_{1-x}Y_xMnO_3$ series as found for $TbMnO_3$ [27]. All these observation suggest that variation of $\langle r_A \rangle$ tunes magnetic interactions in this solid solution in a similar way as found for the $RMnO_3$ compounds. Therefore this system an adequate analogue of the $RMnO_3$ perovskites and facilitates the investigation of how A-type order is destabilized due to structural tuning.

One feature of the $Nd_{1-x}Y_xMnO_3$ phase diagram that differs from that of the $RMnO_3$ manganites is the observation of a magnetic phase separation between the A-type and spin density wave phases. Such behavior suggest that the transition between these two ground states is not continuous but rather it is separated by 1st order like phase boundaries. The observation of this phase separated region is not predicted for the $RMnO_3$ manganites based on Monte Carlo simulations[48, 91] and will be a topic in the following Chapters.

Results of Physical Property Measurements

Chapter 5
Results of Physical Property Measurements

This chapter will show an overview of the magnetoelectric phase diagrams of the $Nd_{1-x}Y_xMnO_3$ series which was determined from measurements of the dielectric constant, electric polarization and magnetization.

Single crystals of $x = 0.40$, 0.45, 0.50 and 0.55 oriented with Laue x-ray diffraction and cut into plates with the plates whose faces were perpendicular to the principal crystallographic axes a, b and c and the typcal dimensions of $6 \times 6 \times 1.5$ mm^3. For the measurements of the dielectric constant ϵ and the polarization P, the plates were polished and sputtered with a thin layer of gold on each side. The samples were mounted in a measurement device as shown in Figure A.3 that can connect both sides to different measurement devices like a voltmeter or an electrometer and can be used in a PPMS. The PPMS supplied the measurement conditions of low temperatures and high magnetic fields.

ϵ was measured at 1 kHz using a Andeen-Hagerling capacitance bridge AH2700. The polarization was obtained via measuring the pyroelectric current as described in Section 2.6. The voltage of 150 V was applied on the sample when cooled down to 2.5 K without an external magnetic field and a magnetic field from 0 T to 14 T was applied afterward. Then the voltage was removed, the plate surfaces were connected once and a Keithley 6517 system electrometer was connected measuring the electrical current that is flowing between the surfaces of the plates while increasing the temperature at a rate of 2 $\frac{K}{min}$ or the magnetic field was swept at a rate of 100 $\frac{Oe}{min}$. P was then calculated by integrating the pyroelectric or magentoelectric current as a function of temperature or field.

Measurements on polycrystalline samples proved the transition from a collinear A-type antiferromagnetic order to an incommensurate spin order with increasing x. Although the powder diffraction results could not answer the exact order of the spins in the incommensurate state, the structural distortion leading to a spin spiral mimics the one of $RMnO_3$ and indicate a ferroelectric ground state for high doped samples.

Polarization measurements on multiferroic $TbMnO_3$ and $DyMnO_3$ show properties that are characteristic for these structures that shall be shortly introduced here, exemplary for $DyMnO_3$[29] whose magnetoelectric phase diagram is shown in Figure 5.1. In $DyMnO_3$ the crystal first show an A-type antiferromagnetic order below T_N. Below the ordering temperature of the cycloidal structure T_s the crystal exhibits a polarization parallel to the

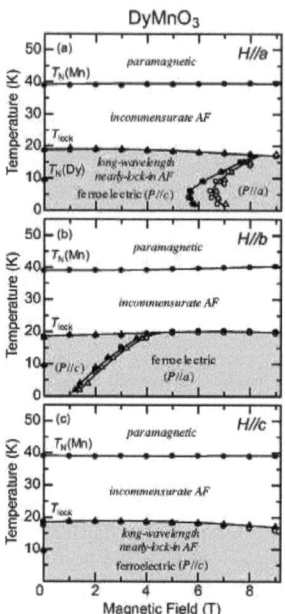

Figure 5.1: Magnetoelectric phase diagram of DyMnO$_3$ with magnetic field along the a, b and c axes. Gray regions indicate ferroelectric phases. Figure published by Kimura et al.[29]

c axis ($P\|c \approx 600\ \frac{\mu C}{m^2}$ at zero field) at 2 K as shown in Figure 5.1. This polarization is not significantly affected by magnetic fields along the c axis up to 9 T (Figure 5.1 (c)). When applying a magnetic field along the a or b axis $P\|c$ is suppressed and the sharp peak of ϵ_c gets diffuse. On the other hand a sharp peak in ϵ_a appears with increasing fields and a $P\|a$ is measurable. This behavior can be connected to a flop of the spin spiral from the bc plane to the ab plane at high fields as it is known from TbMnO$_3$[92, 93] and the multiferroic state may also be excited[94, 95]. The direction of the polarization flops from $P\|c$ to $P\|a$ at $H\|a \approx 6$ T (2 K) (5.1 (a)). A magnetic field along the b axis needs even lower values, like 1 T at 2 K (5.1 (b)). TbMnO$_3$ is to some degree similar but more complex as a high magnetic field along the c axis (like 10 T at 10 K) is able to quench the ferroelectric order into the paraelectric state.

5.1 Systematic Changes of Ferroelectric Properties with x

First the systematic changes of dielectric and ferroelectric properties with x for the samples $x = 0.40$ to 0.55 will be presented. The ϵ_c at $H = 0$ T increases with x as shown in Figure 5.2. The sample $x = 0.45$ is the first one to show a small peak at T_s that is also broader than the ones from higher x samples. The intensity of the peak increases with x and sharpens. T_s is between 24.5 K for $x = 0.45$ and 0.50 and 23.2 K for $x = 0.55$, indicative of a ferroelectric transition.

The temperature dependence of ϵ is in agreement and closely tracks the $P\|c$ data. $P\|c$ at 2 K increases with x starting with $x = 0.45$. The stability of $P\|c$ on applied magnetic field $H\|c$ is also higher with x as shown in Figure 5.3. The field necessary to suppress the polarization increases with x from 4.5 T to approximately 8 T in each direction of the field. For $x = 0.45$ a field of $H = 14$ T is applied it quenches the ferroelectric order and the information is lost when retracting the field to 0 T, no $P\|c$ appears. On the other hand at $x = 0.55$ the polarization is completely regained after applying $H\|c = 14$ T and to 20% with $H\|b$. This indicates that the stability of the ferroelectric order and therefore also of the magnetic order increases smoothly with x instead passing a transition area that exhibits a lower degree of stability.

When the $P\|c$ of the sample $x = 0.55$ is suppressed under a field of $H\|b$, $P\|a$ reaches only a value of 15% of $P\|c$ at zero field and the flop happens at approximately 7 T. On releasing the field down to zero $P\|c$ reaches only 19% of its original value, so this process of spin flop is not reversible with $H\|b$.

The field sweep of $H\|c$ shows a more interesting behavior in the polarization profiles with x as it is displayed in Figure 5.4. Sample $x = 0.55$ shows a flop of the polarization from $P\|c$ to $P\|a$ with $\frac{P\|a}{P\|c} = 47\%$ at 7 T with a full recovery of $P\|c$ when the field is withdrawn and no remanent polarization in $P\|a$. When x is decreased first the recovery of $P\|c$ is affected as it only regains $\frac{P\|a}{P\|c} = 39\%$ at $x = 0.50$ and with $x = 0.45$ no $P\|c$ is recovered and no $P\|a$ appears. This indicates that with lower doping the stability of the ferroelectric under field and is comprimized and quenched at high fields. As sample $x = 0.40$ does not show a polarization with $H\|c$ this measurement is excluded here.

The temperature profiles of $P\|a$ and $P\|c$ under $H\|c$ shows at the transition region of $x = 0.50$ and 0.45 another transition at approximately 8 T that causes an amplification of the polarization. This is quite unusual as transitions below T_s in $DyMnO_3$ and $TbMnO_3$ decreases the polarization if not suppress it completely. On the other hand it can be seen as a decreasing of the polarization and the magnetic order with increasing temperature indicating an unstable component in the magnetic order.

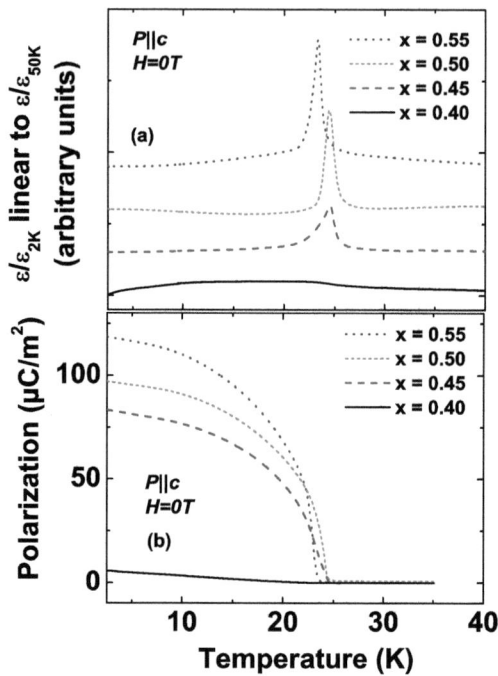

Figure 5.2: Temperature profiles samples $x = 0.40$ to 0.55 of dielectric constant (top, waterfall plot) and polarization (bottom) measured along the c axis at $H = 0$ T and 1 kHz. The values of the dielectric constant are normalized to a linear fit from $\frac{\epsilon}{\epsilon_{2K}}$ to $\frac{\epsilon}{\epsilon_{50K}}$ and the error of the temperature is ± 0.4 K. The error of the polarization is $\pm 1.5 \frac{\mu C}{m^2}$.

5.2 Results of Measurements on $x = 0.55$ sample

The complete results of the measurements for $x = 0.55$ are shown in Figures B.2 and B.6. Like DyMnO$_3$ $x = 0.55$ does not show significant changes along the b direction of the dielectric constant with temperature or field, except a small increasing of the value with $H \| b$ at T_s, and does not have a significant polarization that can be measured in this

5.2. Results of Measurements on $x = 0.55$ sample

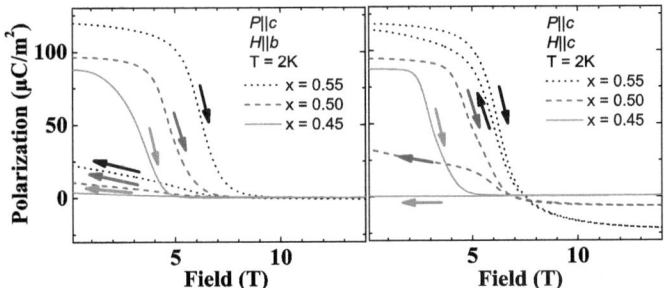

Figure 5.3: Magnetic field profiles of the samples $x = 0.45$ to 0.55 of polarization measured along the c axis with field sweep $H\|b$ (left) and $H\|c$ (right) from 0 T to 14 T and back to 0 T with 100 $\frac{Oe}{min}$ at 2 K.

direction. Applying a magnetic field along the a direction does not change the shape of the temperature profiles of the dielectric constant or the electric polarization.

Remarkable changes in the temperature profiles are observed when a magnetic field is applied along the b and c direction. First the results of the measurements of ϵ_a and ϵ_c with $H\|b$ and $H\|c$ will be discussed and are shown in Figure 5.6. When $H\|b$ or $H\|c$ is applied, a sharp peak in ϵ_c at T_s broadens and decreases in temperature and is completely suppressed at fields over 8 T ($H\|b$) or 5 T ($H\|c$). It should be marked that in contrast to DyMnO$_3$ no second peak below T_s in ϵ_c develops. This peak can be assigned to the ordering of the Dy spins comparable to the ordering of the Tb spins in TbMnO$_3$ as the amount of spins at this crystallographic position is decreased by using Nd instead of Dy and occupied by 45 per cent with nonmagnetic Y this missing peak is explainable. The temperature of this peak will will be labeled T_{Dy}.

The measurement of ϵ_a without an applied magnetic field only shows a kink at $T_s = 23$ K. When applying $H\|b$ the kink sharpens and transforms into a peak. The peak temperature decreases to the lowest value of 15 K at $H = 7$ T while getting stronger and developing to a intense but not too sharp peak at 21.5 K at $H = 14$ T. When applying $H\|c$ the kink of ϵ_a changes to a hump with its maximum changing in temperature with H but stays more or less stable at fields of $H\|c \geq 5$ T with a maximum at 18.5 K. The baseline value of ϵ_a is much smaller than ϵ_c, what is approximately also the case for the $x = 0.50$ sample. This is different to the values for DyMnO$_3$ as the baseline of ϵ_c in both compounds are approximately in the same region (20 for $x = 0.55$ and 17 for DyMnO$_3$)

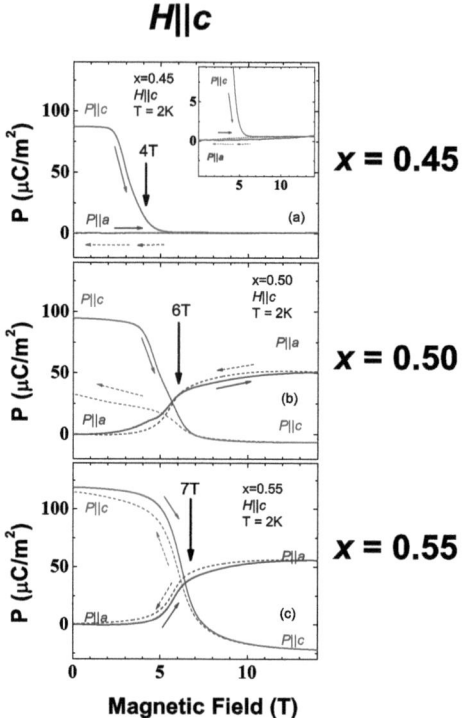

Figure 5.4: *Magnetic field profiles of the samples $x = 0.45$ to 0.55 of polarization measured along the c axis with field sweep $H\|c$ from 0 T to 14 T and back to 0 T with 100 $\frac{Oe}{min}$ at 2 K. Straight lines are increasing field, dashed lines are decreasing field. The inset at $x = 0.45$ shows the graph with magnified y scale. The error of the polarization is $\pm 1.5 \frac{\mu C}{m^2}$.*

but the baseline for ϵ_a are completely different as ϵ_a ($x = 0.55$) = 2.1 and ϵ_a (DyMnO$_3$) ≈ 30 with a very large peak at T$_s$ ($H\|b$) that is nearly 10 times higher.

The measurements of $P\|a$ and $P\|c$ with applying $H\|b$ and $H\|c$ are shown in Figure 5.7. Due to the assumed missing ordering of the A-site ions the zero field polarization with changing temperatures at the $x = 0.55$ sample shows only one second order transition over the complete temperature range. According to the changes in ϵ_b when applying $H\|b$ the onset temperature of $P\|c$ decreases from 23.5 K to 15.5 K at $H\|b = 7$ T while the value of $P\|c$ also decreases from $120\frac{\mu C}{m^2}$, a remarkable lower value compared to the $600\frac{\mu C}{m^2}$ of DyMnO$_3$, and is completely suppressed at fields of $H\|b \geq 8$ T. At DyMnO$_3$ also the polarization with higher fields show that the area between T_{Dy} and T$_s$ shows higher polarization before it is completely suppressed also at $H\|a \geq 8$ T or $H\|b \geq 4$ T.

When applying $H\|c$ T$_s$ of $P\|c$ first decreases to 14 K at $H = 5$ T and increases again at higher fields. Further $P\|c$ is not quenched with increasing field like when $H\|b$ is applied but is inverted. At $H\|c = 6$ T a part of the temperature dependence curve is negative[1]. At $H\|c = 8$ T the whole curve is negative and at $H\|c = 14$ T $P\|c$ is about -23 $\frac{\mu C}{m^2}$ that is 19% of the value at zero field.

On the other hand when $H\|b \geq 6$ T or $H\|c \geq 2$ T is applied a remarkable increasing of $P\|a$ is measurable. The onset temperature of $P\|a$ follows the maximum of ϵ_a. Extraordinary is the fact that $P\|a$ shows a maximum at $H\|b$ that changes with increasing field. At $H\|b = 14$ T $P\|a(2.5K) = 51$ $\frac{\mu C}{m^2}$ but has its maximum at 14.5 K with $P\|a(14.5K) = 128$ $\frac{\mu C}{m^2}$. At fields of $H\|a \geq 5$ T a significant $P\|a$ is measurable the increases in intensity but changes only little at $H\|a \leq 8$ T to a maximum of $P\|a = 65.5$ $\frac{\mu C}{m^2}$ with the transition temperature of 18.5 K. It should be mentioned that these curves are comparable to the results for DyMnO$_3$ whose $H\|b$ curve matches to some degree the $H\|c$ curve measured here and the $H\|a$ of DyMnO$_3$ matches this $H\|b$ curve.

Regarding those results it can be stated that by applying a sufficiently large magnetic field along the transition of the peaks in ϵ_c and ϵ_a from the c axis to the a axis at T$_s$ can be ascribed to the switching of the polarization from $P\|c$ to $P\|a$ as it is visible in Figure 5.7. This is comparable to DyMnO$_3$ that exhibits a polarization flop at approximately $H\|a \approx$ 7 K or $H\|b \approx 3$ K. When doing a field sweep at 2 K this assumption is confirmed as shown in Figure 5.8. At $H\|c$ of approximately 6 T $P\|c$ shifts to $P\|a$, but with only half of the value for $P\|c$. When the magnetic field is withdrawn this flop is nearly completely reversible. At $H\|b$ of approximately 6 T $P\|c$ starts to quench, and $P\|a$ increases only little. When the magnetic field is withdrawn $P\|a$ goes zero again but $P\|c$ only gains 20 per cent of the starting value at zero field, leading to the conclusion that only an applied field along the c-direction is able to induce a reversible flop of the polarization.

---

[1] Negative values of the polarization are always meant to be negative relative to the measured polarization at $H = 0$ T in the context of this work. Negative values therefore mean that the direction of polarization is reversed from $+c$ to $-c$.

5.3 Similarities and Differences to other RMnO$_3$

The results in the measurements of the physical properties show a lot of similarities between the Nd$_{1-x}$Y$_x$MnO$_3$ system and the RMnO$_3$ compounds. The compositions with $x \leq 0.40$ are not ferroelectric and, exemplary for $x = 0.40$, do not show a polarization in any principle crystallographic direction under fields up to 14 T applied in any direction indicating a ground state that is not ferroelectric. This behavior is known from EuMnO$_3$ that does not show a magneto electric effect. The results for $x = 0.45$ and 0.50 show an evolving polarization, getting more stable with higher x under high magnetic fields. The composition with $x = 0.55$ show a stable and reversible polarization. The magnetoelectric phase diagrams for an applied field along all three principle crystallographic axis a, b, and c are displayed Figure 5.9.

The sample with $x = 0.55$ is the single crystal sample with the highest crystallographic distortion and the systematic changes of the ferroelectric properties with x indicate that this sample is the one that is most similar to the multiferroic RMnO$_3$, especially to DyMnO$_3$. Both show a $P\|c$ below a T$_s$ as a result of a cycloidal spin order with the spiral in the bc plane and the propagation direction along the b-direction. In both crystals $P\|c$ can be flopped to $P\|a$ by an external magnetic field. A characteristic difference is that this flop is strongest with $H\|b$ in DyMnO$_3$ and $H\|c$ has no effect while in $x = 0.55$ the polarization flop with $H\|c$ is strongest and even reversible when the field is withdrawn while $H\|a$ shows no effect.

Another significant difference is the absence of a second transition in the polarization order below T$_s$ under zero fields. This second transition is caused by the onset of the magnetic order of the rare earth ion. A magnetic order of Nd influencing the Mn order, like Dy or Tb in RMnO$_3$, could not be observed while Y is completely non-magnetic, so the results of the polarization measurements are only the result of the Mn spin order.

So it can be shown that changes in x systematically changes the magnetic and ferroelectric properties of the system. With increasing x first a $P\|c$ appears that can be quenched with magnetic fields strong enough of $H\|b$ and $H\|c$. When $P\|c$ is stable enough the polarization can be flopped to some degree to $P\|a$ under $H\|b$ and $H\|c$. At $x = 0.55$ $P\|c$ is not quenched under high fields but is reversed to some degree. The flop of the polarization can be explained by the flop of the cycloidal plane as it is known for TbMnO$_3$[93].

5.3. Similarities and Differences to other $RMnO_3$

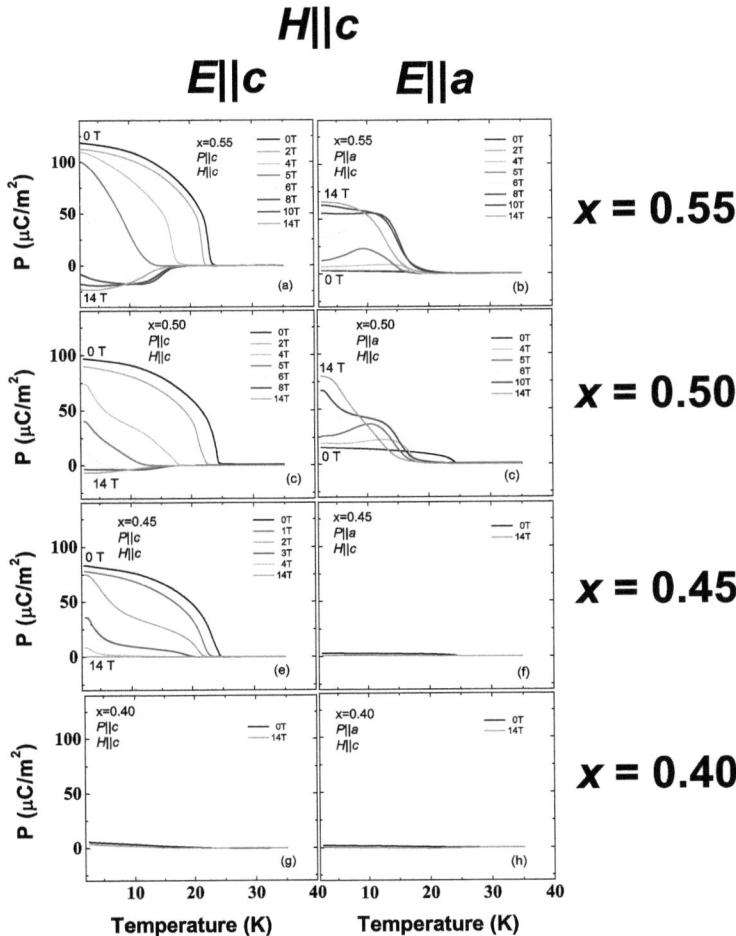

Figure 5.5: Temperature profiles of samples $x = 0.40$ to 0.55 of polarization along the a and c axis at magnetic fields up to $H = 14$ T along the c axis.

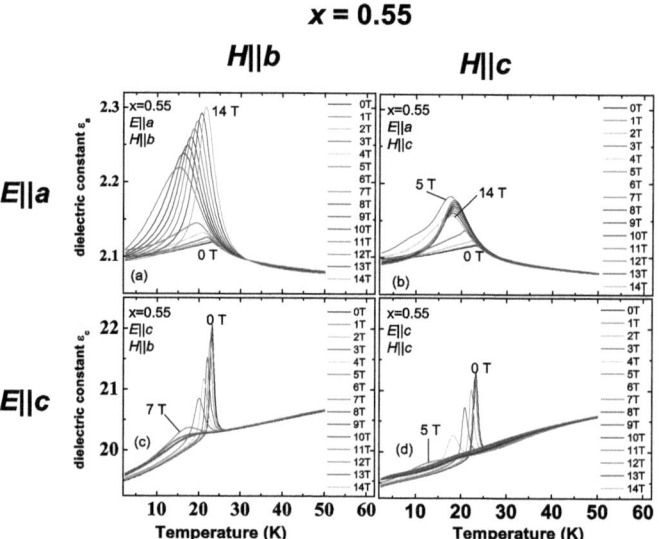

Figure 5.6: Sample $x = 0.55$ temperature profiles of dielectric constant along the a axis (a)-(b) and the c axis (c)-(d) at magnetic fields up to $H = 14$ T along the b and c axis.

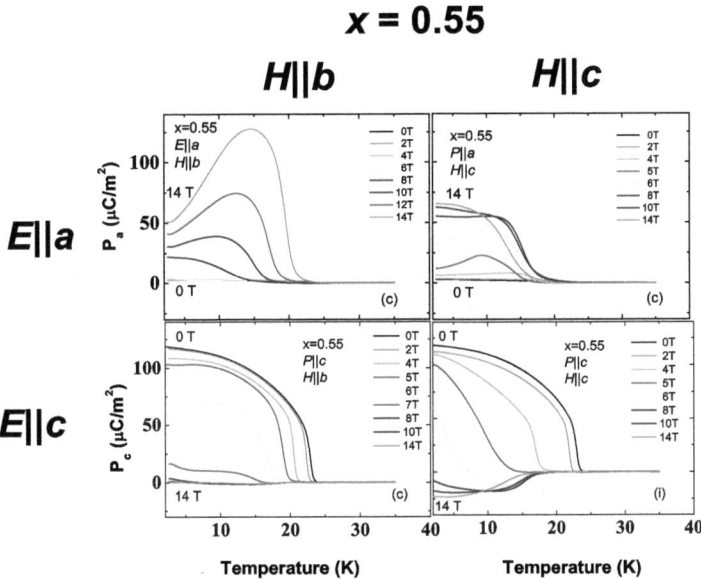

Figure 5.7: Sample $x = 0.55$ temperature profiles of polarization along the a axis (a)-(b) and the c axis (c)-(d) at magnetic fields up to $H = 14$ T along the b and c axis.

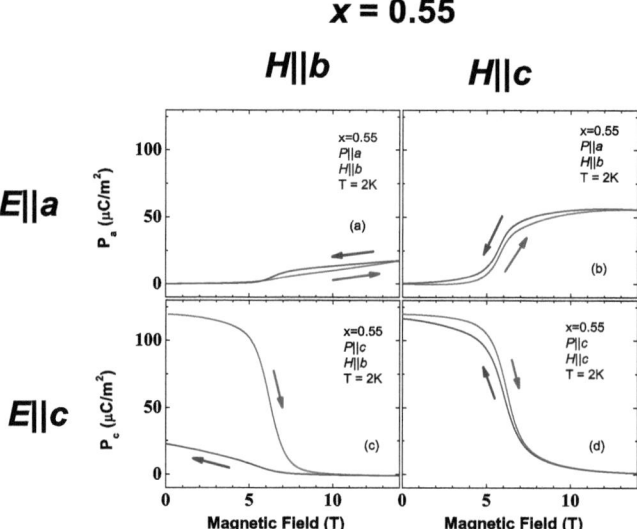

Figure 5.8: Sample $x = 0.55$ magnetic field profiles of polarization along the a axis (a)-(b) and the c axis (c)-(d) at magnetic field sweep from $H = 0$ T up to 14 T and back to 0 T with $H \| a$ and $H \| a$.

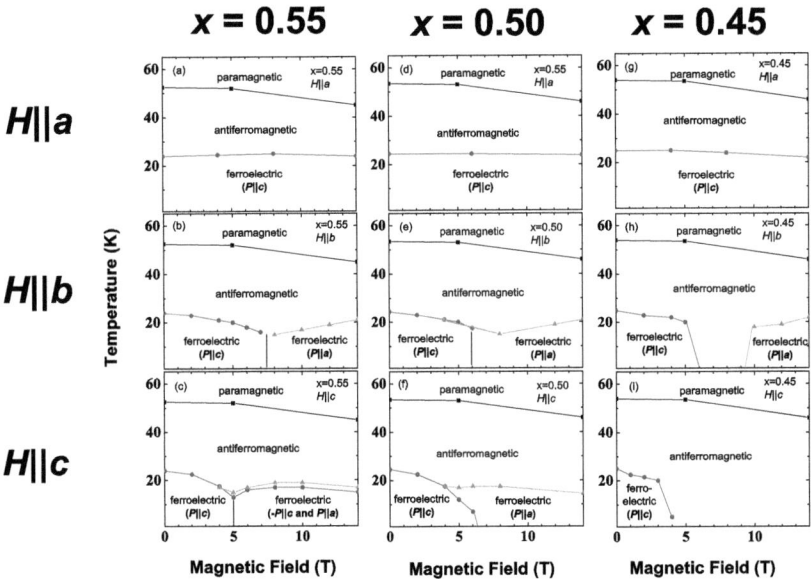

Figure 5.9: *Magnetoelectric phase diagram of the $x = 0.45$ to 0.55 samples with magnetic fields along the a, b, and c axes. T_N was obtained with magnetization measurements as described in Section 4.1 that were confirmed at $H = 0$ T with single crystal neutron diffraction. The data obtained by measurements of dielectric constant, pyroelectric and magnetoelectric current is displayed in circles and triangles.*

Chapter 6
Magnetic Order in $Nd_{1-x}Y_xMnO_3$

Polycrystalline sample measurements revealed a transition from A-type antiferromagnetic order to an incommensurate spin density wave with a transition zone where both magnetic components co-exist. The measurements of the physical properties proved that the compounds with $x \geq 0.45$, having a dominant incommensurate spin component, show a polarization along the c-direction below T_s. To examine in detail the development of the magnetic structure with varying x and temperature, neutron diffraction measurements on single crystals are necessary.

Single crystals were grown of the compositions $x = 0.3$ to 0.55 as described in Section 3.2. Although it was possible to grow single phase $x = 0.6$ polycrystalline samples it was not possible to grow single crystals as either the growth resulted in a polycrystalline sample or split up into a Nd-rich orthorhombic phase and a Y-rich hexagonal phase. Neutron diffraction measurements were conducted at the Helmholtz-Zentrum Berlin für Materialien und Energie at E4, E5 and V2 and at Institut Laue-Langevin at D10, for more details to this instruments see Section A. All temperature dependent measurements were done on measuring at heating up with distinct temperature steps in contrast to the measurements with polycrystalline samples at D20 that were done with a temperature sweep. The crystals grown including the instruments they were measured with is listed in Table 3.2 on page 35.

6.1 Magnetization Measurements on Single Crystals

Temperature dependent magnetic susceptibility measurements were conducted in a, b and c direction at $H = 500$ Oe[1] as shown in Figure 6.1. Those measurements confirm the results of the magnetization measurements with the polycrystalline samples and of the physical properties measurements in Chapter 5. The direction dependent measurements can clearly distinguish between the different transitions. The antiferromagnetic transition is in the b-direction that is the easy axis confirms the results in Section 7.2 that the collinear spin

---

[1] Also measurements at $H = 5$ T and 14 T were conducted and the whole results are displayed in Figure B.1

density wave align along the b-direction. T_N here is about 2 K higher than measured with polycrystalline samples. The ferromagnetic transition that was also measured at the polycrystalline samples decrease with x and is not affecting the measurement anymore at $x \leq 0.50$ and is strongest along $H\|c$ but also to some degree along the other directions. T_N of the A-type antiferromagnetic order also decreases with x while the T_N of the incommensurate component remains at approximately 50 K as it will be shown further in this chapter. This confirms the assumption that this order is assigned to the ordering of the Nd^{3+} spins along the c direction when the Mn^{3+} order in an A-type antiferromagnetic orientation as described in Section 4.1.

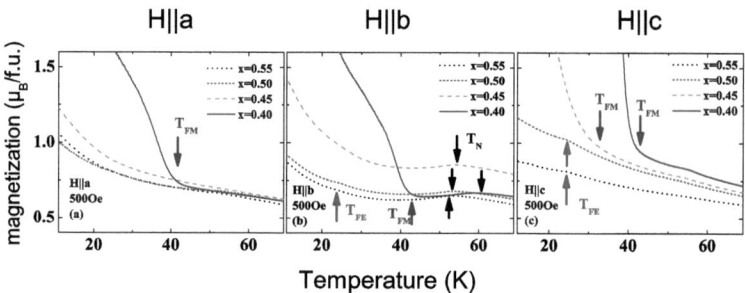

Figure 6.1: Magnetization as a function of temperature for the samples $x = 0.4$ to 0.55 along the a, b, and c axis at a magnetic field of $H = 500$ Oe. Arrows show T_N, the ferromagentic transition T_{FM} and the ferroelectric transition temperatures T_{FE}.

The evolution of the ferroelectric phase is visible at $H\|b$ and $H\|c$ at samples $x = 0.50$ and 0.55 that confirms the results from Section 7.2 that a magnetic transition takes place in the b and c axis by transforming the collinear spin density wave along the b axis to the cycloidal spin structure in the bc plane propagating along the b axis.

6.2 Variation of Magnetic Order with x

One result of the measurements on the polycrystalline samples was the observation of the co-existence of the A-type antiferromagnetic order and a modulated spin order. But as the value of 2Θ of the magnetic components are quite alike, the q-resolution of the experiment did not allow for a closer investigation. In this Section temperature dependent neutron diffraction scans along the (0k1) direction will be shown that cover both magnetic reflections and will also provide the wavenumber of the incommensurate reflection.

6.2. Variation of Magnetic Order with x

The results of q-scans along the (0k1) direction of the samples $x = 0.35$ to 0.50 are displayed in Figure 6.2 and confirm the co-existence of the A-type and the incommensurate order in the compounds $x = 0.40$ and 0.45. The instrumental and sample resolution allows to resolve the two different magnetic reflections and measure their development with x. The low x compounds show only a pure A-type antiferromagnetic order while on the other hand the samples $x \geq 0.50$ showed only a modulated spin component at 2 K. The wavenumber of the incommensurate component is increasing with x. The compounds with $x = 0.35$ to 0.45 clearly show both magnetic components and will be called in the text below the **transition zone**. The full width half maximum of the magnetic reflections in the transition zone is higher than the instrumental resolution indicating a smaller coherence length.

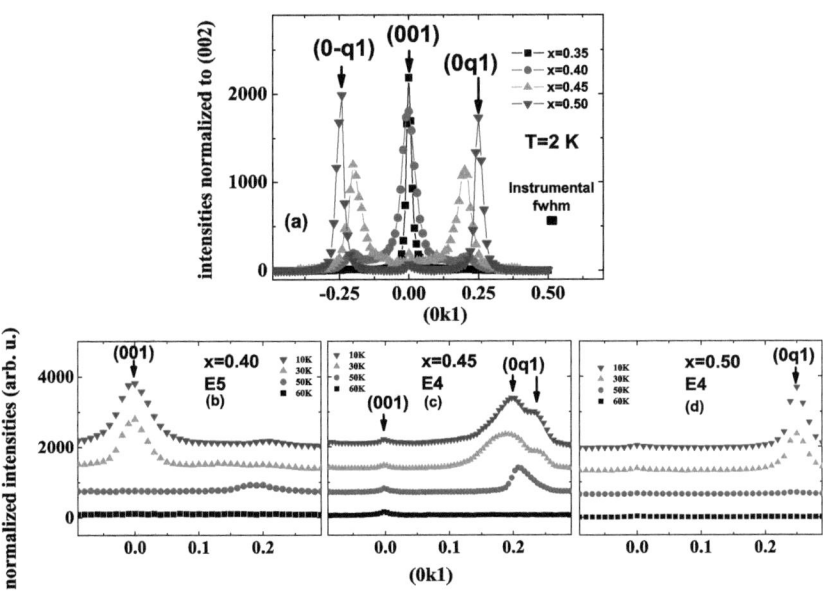

Figure 6.2: *Neutron diffraction q-scans measured from samples $x = 0.35$ to 0.50 from (0 -0.5 1) to (0 0.5 1) (a) with the intensities normalized to the value of the (002) reflection at 2K. The panels (b) to (d) show temperature dependent waterfall plots of q-scans from (0 -0.1 1) to (0 0.3 1) for the samples $x = 0.4$, 0.45 and 0.5 with a separation of the plots of about 2000 units per temperature.*

A closer look at the temperature dependence of the reflections in the transition zone (Figure 6.2 (b)-(d)) reveals that the onset temperature of the A-type antiferromagnetic order is lower than T_N of the magnetic susceptibility measurements. The magnetic measurements confirm those transition temperatures as shown in Section 6.1. Further not only the shape of the satellite reflection of the $x = 0.45$ compound is changing with temperature (in contrast to the component without an A-type antiferromagnetic component like $x = 0.50$) but also more than one peak is visible.

The development of the magnetic reflection with doping can also be visualized by the color plots as shown in Figure 6.3. The intensity in the plots is scaled logarithmically to make also the weaker magnetic reflections evident. Here it is very clear to see how in addition to the A-type antiferromagnetic component, a second incommensurate magnetic component, appears and with increasing x it co-exists with the A-type order and eventually replaces it for $x > 0.50$. and for sufficiently larger values of x, it replaces the A-type component.

When comparing the results of the susceptibility measurements with the results of the temperature dependent neutron diffraction measurements (Figure 6.4) it is visible that the onset of the A-type antiferromagneticorder induce the ferromagnetic order along the c-direction of Nd^{3+} spins. In the transition zone ($x = 0.40$ and 0.45), T_N and the intensity of the A-type antiferromagnetic component at 2 K decreases quickly (bottom panel). Further the onset of the A-type antiferromagnetic component in the transition zone also decreases fast with x and is not measurable anymore in $x \geq 0.50$. This is peculiar as the measured T_N in the susceptibility measurements is not lower than 50 K. By comparing the onset of the A-type antiferromagnetic order with the magnetization measurements it is visible that the onset T_N (A-type) is the same as the onset of the ferromagnetic component T_C (Nd) determined in the anisotropic susceptibility measurements. This observation confirms the assumptions made in Section 2.4 that the A-type antiferromagnetic order exposes a small ferromagnetic component along the c-axis due to the Dzyaloshinskii-Moriya interaction of the Mn spins. This induces the Nd spins to align according to this ferromagnetic component resulting in a stronger ferromagnetic moment as described elsewhere for $NdMnO_3$[83, 85].

T_N of the magnetic susceptibility measurements of the compositions of $x \geq 0.40$ is therefore referred to the onset of the modulated spin component and not to the A-type antiferromagnetic component that is a strong indicator to phase separation into two magnetic components.

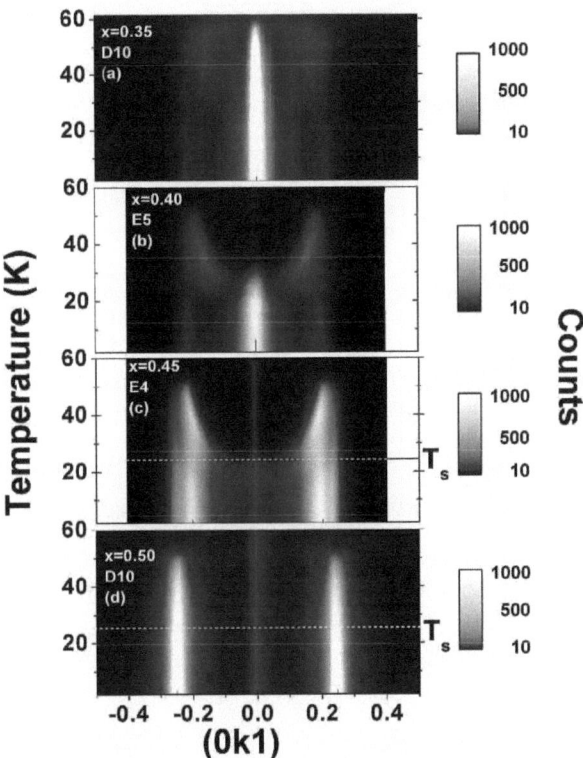

Figure 6.3: *Temperature dependence of neutron diffraction q-scans along the b-axis from (0 -0.5 1) to (0 0.5 1) for the samples $x = 0.35$ (a) to 0.50 (d). The temperature scale is logarithmic. The temperature independent fine line at (001) is the $\frac{\lambda}{2}$ reflection of the nuclear (002) reflection.*

6.3 Magnetic Order in low Doped Region ($x = 0.30$, 0.35)

The sample $x = 0.3$ showed purely the A-type antiferromagnetic reflections. The manganese spins in the $x = 0.35$ sample order nearly exclusively in an A-type antiferromagnetic

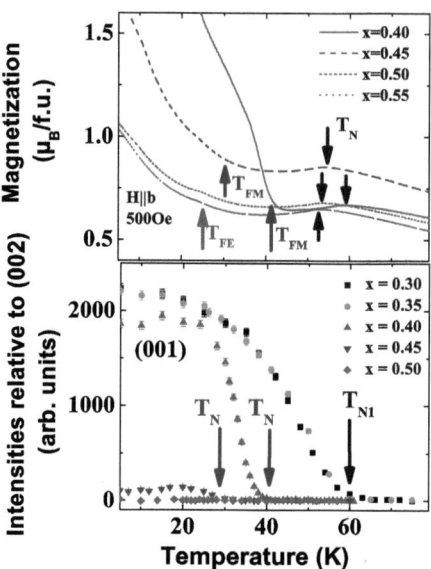

Figure 6.4: *Temperature dependence of the magnetization for $H\|b = 500$ Oe (top) and of the intensities of the A-type antiferromagnetic peak at (001) with the intensity relative to the nuclear (002) reflection (bottom).*

order, even this sample showed the presence of a small diffuse incommensurate reflection, q' with a wavenumber of approximately 0.21 (Figure 6.5). This confirms the results from polycrystalline measurements where the samples of $x \geq 0.35$ displayed only one A-type peak (as displayed in Figure 4.7) and an additional incommensurate component did not improve the Rietveld refinements due to the lower sensitivity of the polycrystalline measurement compared to single crystal measurements.

This incommensurate peak can be fitted with a single gaussian peak as it is displayed in Figure 6.6 giving a coherence length at 8.3 Å along the b-direction that is not even two unit cells and therefore this magnetic component arises from local spin order. When the sample is cooled down the A-type and the incommensurate reflection appear at approximately the same temperature as shown in Figure 6.7. While the intensity of the A-type reflection increases, the incommensurate peak first increases then decreases in intensity.

6.4. Magnetic Order in Phase Co-Existence Region ($x = 0.40, 0.45$)

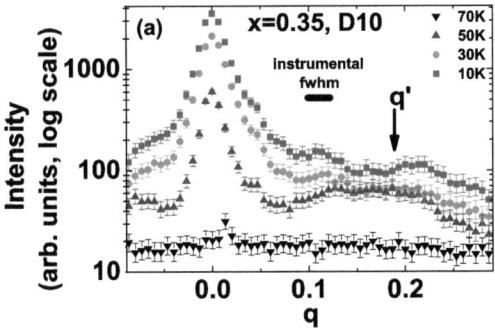

Figure 6.5: *Q scan of sample $x = 0.35$ from (0 -0.09 1) to (0 0.29 1) measured at D10 at different temperatures with a logarithmic y-axis scale.*

The wavenumber increases until the A-type reaches its saturation value. The incommensurate peak remains constant in intensity and wavenumber. This suggests that the two magnetic orders are coupled as they both show the same T_N.

6.4 Magnetic Order in Phase Co-Existence Region ($x = 0.40, 0.45$)

6.4.1 The $x = 0.40$ region

In the transition zone the A-type antiferromagnetic component co-exists with the modulated spin order at 2 K[2]. The q-scans along (0k1) at different temperatures are shown in Figure 6.3 (b) and (c), reveal that T_N of the A-type antiferromagnetic component is not only lower than T_N of the modulated spin component but the position and shape of the satellites change significantly with temperature. The reflection of the modulated spin component q' in $x = 0.40$ at 2 K can be fitted with only one gaussian, as displayed in Figure 6.8(a) giving a coherence length of 15.4 Å along the b-direction that is approximately three unit cells and therefore this magnetic component arises from local order. However on heating a second peak, q", appears at 26 K and can be fitted separately from q' (Fig. 6.8(b)). Above 47 K q' disappears and q" is the only incommensurate peak to be measured. Both q' and q" are broader than the instrumental FWHM resolution and are

---

[2] The analysis and the results of the spin structure is discussed in Section 7

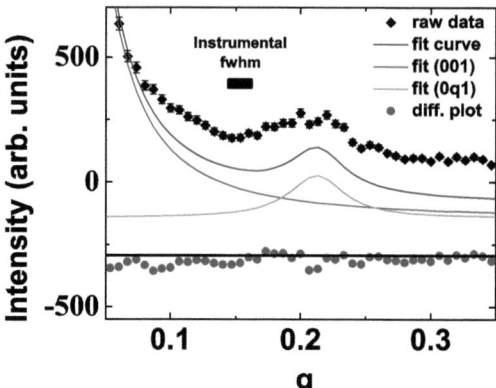

Figure 6.6: Modeling of the incommensurate magnetic reflection of the sample $x = 0.35$ performed on q-scan from (0 0.05 1) to (0 0.35 1) with one peak. The graph shows the fit of the (0q1) reflection, the tail of the fit for the strong (001) reflection and the sum of those two. The cure fits have an offset of -150 units while the difference plot has an offset of -300 units for better visualization.

comparable to the incommensurate peak of the $x = 0.35$ sample[3], in terms of the peak-shape and position change with temperature, but they have a longer coherence along the b-direction.

To have a closer look, the behavior of the magnetic reflections with temperature are displayed in Figure 6.9 for the $x = 0.4$ sample. The different transition temperatures are evident from the temperature dependence. On cooling the q" reflection appears at T_N with increasing intensity and decreasing wavenumber. At the onset of the A-type peak the intensity of q" decreases while a new q' appears with a constant wavenumber over the whole temperature range. When the A-type peak reaches its saturation value q" disappears and the FWHM of the q' peak remains constant. Both the A-type and the incommensurate reflections have a coherence length of approximately three unit cells along the b-direction with 17.9±0.6 Å or 15.4±0.5 Å, respectively.

---

[3]In the following text q' and q" is used to distinguish two comparable incommensurate reflections of the same crystal (either $x = 0.40$ or 0.45) that have a lower and temperature dependent wavenumber.

6.4. Magnetic Order in Phase Co-Existence Region ($x = 0.40, 0.45$)

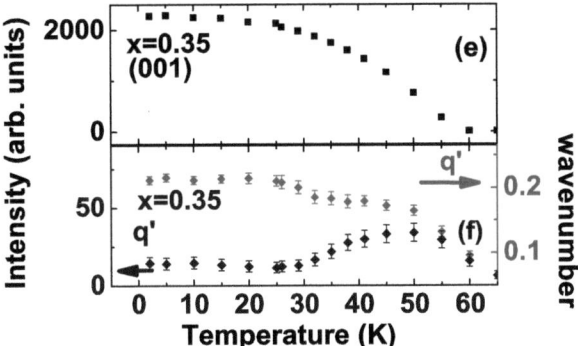

Figure 6.7: *The temperature dependence of the intensity of the magnetic reflections and the wavenumber of the incommensurate q' reflection. These parameters were obtained by the fitting shown in Figure 6.6*

6.4.2 The $x = 0.45$ region

For the $x = 0.45$ sample incommensurate reflections are observed with complex peak shape at 2 K. These are not possible to fit with only one peak. Three peaks are necessary for

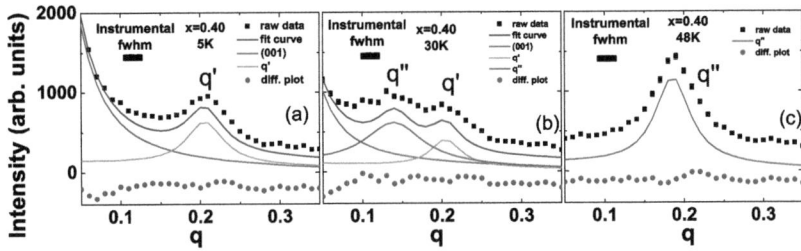

Figure 6.8: *Modeling of the incommensurate magnetic reflection of the sample $x = 0.40$ performed on q-scan from (0 0.05 1) to (0 0.35 1) with one gaussian peak at 5 K (a), 30 K (b) and 48 K (c). It is shown the fit of the (0q1) reflection (light for q', dark for q"), the tail of the fit for the strong (001) reflection and the sum of those two. The peak fits have an offset of -200 units while the difference plot has an offset of -400 units for better visualization.*

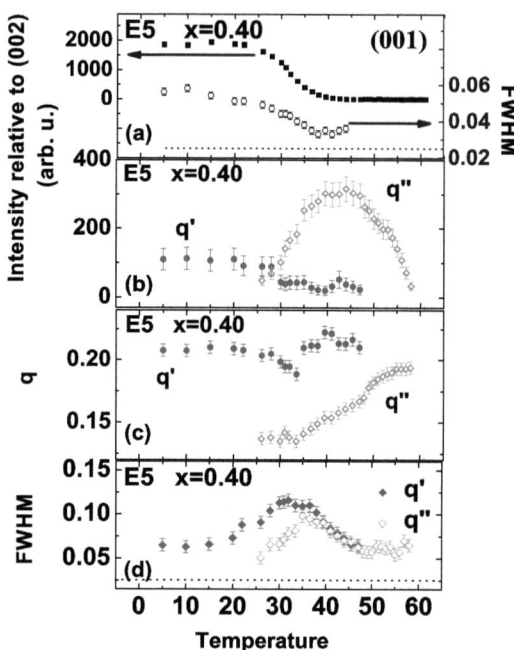

Figure 6.9: Sample $x = 0.40$ temperature dependence of the intensities of the (001) (a) and (0q1) reflections (b), of the wavenumber of the incommensurate reflections (c) and the full width at half maximum for the (0q1) (d) and the (001) (a) reflections measured at E5. The dashed horizontal line in (a) marks the instrumental resolution. These parameters were obtained by the fitting model shown in Figure 6.8.

a good fit but with different FWHM as shown in Figure 6.10, where as in 6.10(b) a fit of the data with gaussian with the same FWHM. The two peaks with lower wavenumber (q' and q") show a temperature dependent wavenumber while the peak with the higher wavenumber (q_{cyc}) behaves like the one of $x \geq 0.50$ with a constant wavenumber with temperature. On the other hand the (001) A-type peak is still measurable. The intensity of the A-type (001) peak can be measured but the peak is too broad to obtain a reliable quantitative fit. Panel 6.11(a) shows only the intensity at the position (0 0.0375 1) that is neither covered by the $\frac{\lambda}{2}$ reflection nor by one of the incommensurate reflections and

6.4. Magnetic Order in Phase Co-Existence Region ($x = 0.40, 0.45$)

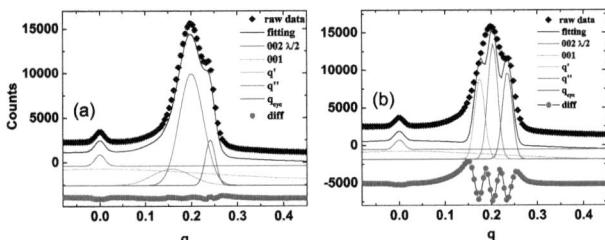

Figure 6.10: *Two possible models of the incommensurate magnetic reflection of the sample $x = 0.45$ performed on q-scans from (0 0.09 1) to (0 0.32 1) with three gaussian of either different (a) of the same (b) fwhm. The main black curve is the sum of the peak fits, the other peaks display the different peaks used. The black curve has an offset of -500 units, the other peaks have an offset of -1000 or -2000 units, while the difference plot has an offset of -5000 (a) or -7000 (b) units for better visualization.*

therefore may only reflect a signal from the A-type diffuse scattering that is independent of a model. This position shows a clear onset temperature at the same temperature as a fitted broad A-type reflection (black) in panel 6.11(b), suggesting the existence of a residue of an A-type component, although the obtained model parameters are not reliable.

The sample $x = 0.45$ occurs at the composition at the very threshold between the two magnetic order A-type, that is barely detectable in this composition, and the incommensurate spin order. On cooling all three incommensurate components appear at T_N and increase in their intensity. At the onset of the A-type order their intensity decreases until T_s, where the intensity increases again. When the A-type component reaches its saturation value, the intensity of the incommensurate components remains constant with decreasing temperature. When looking at the wavenumber only q_{cyc} is invariant with temperature while q' and q'' vary with temperature having a minimum at the onset of the A-type. The FWHM of those peaks suggests that q' arises from short range order (10 ± 1 Å) while q'' has a longer coherence at T_N (16.4 ± 0.6 Å) and q_{cyc} shows the highest coherence with 40 ± 1 Å below T_s.

In both samples of the transition zone it can be stated that the A-type antiferromagnetic component is less stable than the modulated spin component as T_N of the A-type component is lower than T_N of the incommensurate component. Further the evolution of the A-type antiferromagnetic order with decreasing temperature influences the incommensurate components in a way that the intensity and partly also the wavenumber of those

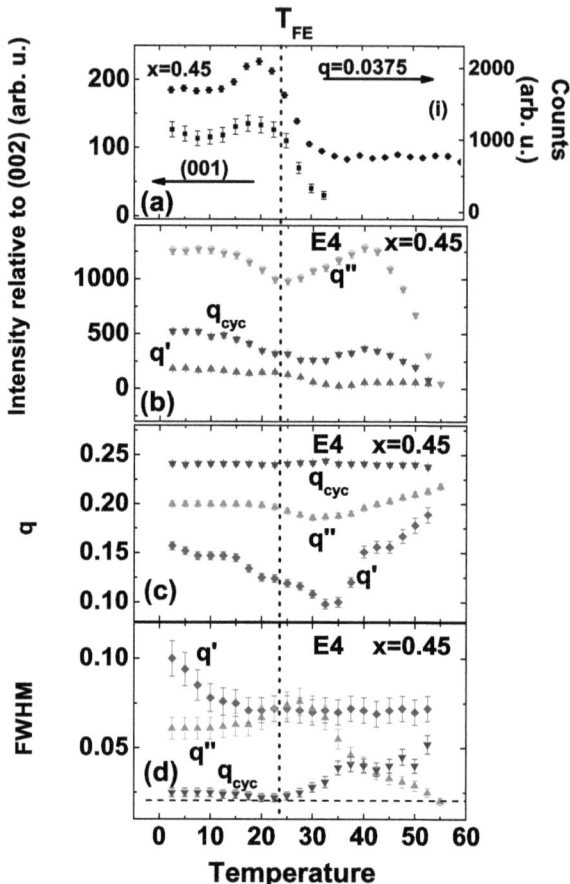

Figure 6.11: Sample $x = 0.45$ temperature dependence of the intensities at the $(0\ 0.0375\ 1)$ position (a), of the (001) and $(0q1)$ reflections (b), of the wavenumber of the incommensurate reflections (c) and the full width at half maximum for the $(0q1)$ (d) reflections measured at E4. The dashed horizontal line in (d) marks the instrumental resolution. These parameters were obtained by the fitting model shown in Figure 6.10.

6.5 Magnetic Order in the Higher Doped Region ($x = 0.50, 0.55$)

varies until the A-type reaches its saturation value.

Samples of $x = 0.50$ and 0.55 do not display an A-type antiferromagnetic component but only a modulated spin component with a wavenumber of $q = 0.250$ and 0.275, respectively. These values are constant with temperature as shown for $x = 0.50$ in Figure 6.3 (d). This is in contrast to the results for TbMnO$_3$ where the wavenumber varies as a function of temperature between T$_N$ and T$_s$[96]. In this behavior Nd$_{1-x}$Y$_x$MnO$_3$ is comparable to Eu$_{1-x}$Y$_x$MnO$_3$ (with non-magnetic Eu and Y) where the wavenumber is also stable for the multiferroic phase with temperature and magnetic field[65], so the influence of the magnetic Nd^{3+} on the magnetic order of the manganese is in Nd$_{1-x}$Y$_x$MnO$_3$ very little compared to Tb^{3+}.

T$_N$ of the incommensurate reflections always match T$_N$ from the magnetic measurements and the magnetic reflections in the q-scans can be fitted for $x = 0.50$ and 0.55 perfectly with only one gaussian peak. Those samples exhibit a spontaneous polarization at approximately 24 K. When the intensity of the magnetic $(0q1)$ reflection is plotted versus temperature a kink is visible in the curve of the reflection and is even clearer for the $(02 - q1)$ reflection as displayed in Figure 6.12. This reflects the magnetic transition at T$_s$ = 24 K from collinear spin density wave with only one magnetic m_y component to cycloidal spin order with a m_y and a m_z component, visible in the temperature dependent intensity of the magnetic reflections.

6.6 Discussion of the Magnetic Order

The transformation of the magnetic ground state from A-type antiferromagnetic order to a spin cycloid with a transition zone at $x = 0.40$ and 0.45 is difficult to interpret. Figure 6.13 displays the variance of the coherence length of the different magnetic components from $x = 0.30$ to 0.55 at 2 K to summarize the results[4]. It is visible that when the (001) reflection indicates local order of the A-type order ($x = 0.40$ to 0.45) and one or two co-existing modulated spin component(s) with a longer wavenumber of approximately $q \approx 0.20$, those components are also short ranged ordered (15 Å counting approximately 3 unit cells along the b-axis). Those components are directly connected to the A-type antiferromagnetic

---

[4]The sample resolution on the respective instrument is determined by the q-resolution of the nuclear (002) peak that is the nuclear reflection in the bc-plane that is closest to the (001) reflection and also one of the strongest peaks.

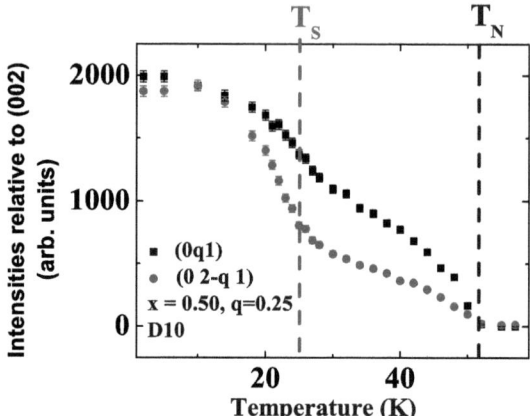

Figure 6.12: Results of neutron scattering of sample $x = 0.50$ showing the temperature dependence of the intensity of the $(0q1)$ and the $(02 - q1)$ reflection.

component as they only exist when the A-type antiferromagnetic component is present at low temperatures and their wavenumber is directly affected by the emergence of the (001) reflection. On the other hand the modulated spin component with a higher wavenumber exhibits at 2 K a coherence length of about the resolution of the respective sample on that instrument. This indicates that the A-type antiferromagnetic order in the transition region is stabilized together with local magnetic spin order with lower wavenumbers of about $q < 0.23$. On the other hand the spin order with wavenumbers of $q > 0.23$ suggest to be responsible for the spontaneous polarization below T_s shown in Chapter 5. Further the wavenumber of the most intense incommensurate reflection in the sample changes linearly with x, as it is also known from $Eu_{1-x}Y_xMnO_3$[65].

The temperature dependent measurements therefore suggest that the transition from the non-multiferroic A-type antiferromagnetic order to the multiferroic cycloidal spin spiral with x is of first order. With increasing x the magnetic frustration increases causing the A-type antiferromagnetic structure to became instable. But the transition to the modulated spin structure takes place piecemeal. In this region T_N of the A-type antiferromagnetic structure decreases and is accompanied by decreasing coherence length with x, while a second and third modulated long wavelength spin order co-exists. This second component is different to the cycloidal spin spiral of $x = 0.50$ and 0.55 for the wavenumber is temperature dependent or, more precisely, appears coupled to the (001) reflection from the A-type

6.6. Discussion of the Magnetic Order

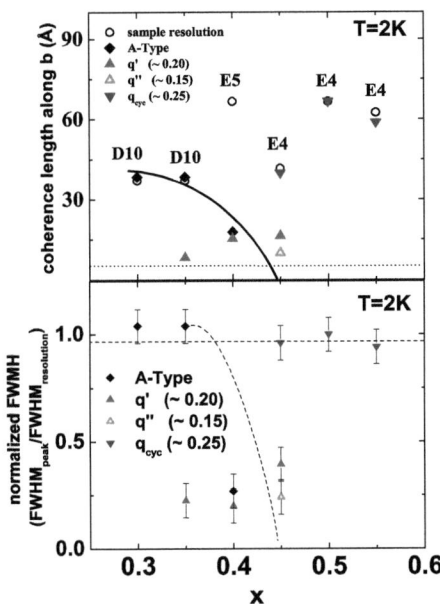

Figure 6.13: *The top panel shows the coherence length depending on composition from sample $x = 0.30$ to 0.55 showing the maximum resolution of the respective samples on the instruments with squares, A-type component (001) with circles, the modulated spin component with low wavenumber of approximately $q \approx 0.20$ in triangles (tip up) and the modulated spin component with higher wavenumber of approximately $q \approx 0.25$ in triangles (tip down) at 2 K. E4 ($x = 0.45$) was made in high flux mode while E4 ($x = 0.50$ and 0.55) measurements are made in high resolution mode, D10 and E5 were used in high flux mode. The dashed horizontal line marks the unit cell size along the b-direction. The bottom panel shows the data on a normalized scale relative to the sample resolution.*

antiferromagnetic component. For sufficiently large x, wavenumbers close to $\frac{1}{4}$ are found together with the magnetic cycloidal order that induces a spontaneous polarization as measured in Chapter 5.

Cycloidal Order in $Nd_{1-x}Y_xMnO_3$

Chapter 7
Cycloidal Order in $Nd_{1-x}Y_xMnO_3$

The solid solution $Nd_{1-x}Y_xMnO_3$ can simulate the general structural and magnetic behavior of the rare earth $RMnO_3$. At low x values the ordering of Mn-spins is antiferromagnetic A-type, while for $x \geq 0.5$ only an incommensurate spin spiral phase is found as the magnetic ground state. For intermediate compositions of $0.4 \leq x \leq 0.45$ a novel region of co-existence between the A-type and spin spiral magnetic ground states exist. The incommensurate spin spiral state shows a ferroelectric polarization along the c-direction. The detailed structure of the spin spiral shall be the focus of this chapter.

The crystals were used on the D10 four axis diffractometer at ILL, Details are displayed in Section A.12. The samples were measured at high flux mode at 2.36 Å at 2 K and the data was analyzed with Racer and LAMP[1] runtime environment Version 6.4.

7.1 The Irreducible Representations of the $Pbnm$ Space Group

The magnetic moment of the atom 'j', in the cell with origin at the lattice point $\mathbf{R}_n$, in a magnetic structure characterized by the set of propagation vectors $\{\mathbf{k}\}$, is given by the Fourier series: $\mathbf{m}_{nj} = \sum_{\{k\}} \mathbf{S}_{\mathbf{k}j} \cdot e^{-2\pi i \mathbf{k} \cdot \mathbf{R}_n}$, where the sum is extended to the whole set of propagation vectors. For $\mathbf{k}$ at the interior of the Brillouin zone the vector $-\mathbf{k}$ should also be present and $\mathbf{S}_{\mathbf{k}j} = \mathbf{S}^*_{-\mathbf{k}j}$. The physical meaning of the basis functions of the irep Γ_ν in describing a magnetic structure is that of the Fourier coefficients $\mathbf{S}_{\mathbf{k}j}$ are linear combinations of the constant vectors $\mathbf{S}_\alpha(\mathbf{k}, \nu, \lambda|j)$, so that

$$\mathbf{S}_{\mathbf{k}j} = \sum_{\alpha\lambda} C_{\alpha\lambda} \mathbf{S}_\alpha(\mathbf{k}, \nu, \lambda|j). \tag{7.1}$$

For a given propagation vector $\mathbf{k}$ and a given irep Γ_ν, the index α runs between 1 and n_ν and λ between 1 and d_ν. The coefficients $C_{\alpha,\lambda}$ can be real or purely imaginary. By varying these coefficients, the whole class of magnetic structures satisfying the symmetry

---
[1]LAMP stands for Large Array Manipulation Program and the LAMP package is distributed at the ILL in the public domain

Γ_ν	1	m_z	2_{1y}	b_z	Mn$_1$	Mn$_2$	Mn$_3$	Mn$_4$	mode
Γ_1	1	1	ω	ω	(100)	(-100)	(-ω00)	(ω00)	A_x
					(010)	(0-10)	(0ω0)	(0-ω0)	G_y
					(001)	(00-1)	(00-ω)	(00-ω)	C_z
Γ_2	1	-1	ω	-ω	(100)	(100)	(-ω00)	(-ω00)	C_x
					(010)	(010)	(0ω0)	(0ω0)	F_y
					(001)	(00-1)	(00-ω)	(00ω)	A_z
Γ_3	1	-1	-ω	-ω	(100)	(-100)	(ω00)	(-ω00)	G_x
					(010)	(0-10)	(0-ω0)	(0ω0)	A_y
					(001)	(001)	(00-ω)	(00ω)	F_z
Γ_4	1	-1	-ω	ω	(100)	(100)	(ω00)	(ω00)	F_x
					(010)	(010)	(0-ω0)	(0-ω0)	C_x
					(001)	(00-1)	(00ω)	(00-ω)	G_x

Γ_ν	1	m_z	2_{1y}	b_z	R_1	R_2	R_3	R_4	mode
Γ_1	1	1	ω	ω	(001)	(00-ω)	(001)	(00-ω)	a_z
Γ_2	1	-1	ω	-ω	(100)	(-ω00)	(100)	(-ω00)	a_x
					(010)	(0ω0)	(010)	(0ω0)	f_y
Γ_3	1	-1	-ω	-ω	(001)	(00ω)	(001)	(00ω)	f_z
Γ_4	1	-1	-ω	ω	(100)	(ω00)	(100)	(ω00)	f_x
					(010)	(0-ω0)	(010)	(0-ω0)	a_y

Table 7.1: *The symmetry analysis for space group Pbnm for the magnetic wavevector $k = q_y b^*$, $q_y < 0.5$. The numbering of atoms in the 4b site is: $Mn_1(\frac{1}{2},0,0)$, $Mn_2(\frac{1}{2},0,\frac{1}{2})$, $Mn_3(0,\frac{1}{2},\frac{1}{2})$, $Mn_1(0,\frac{1}{2},0)$. The rare earth atoms of the first 4c orbits are: $R_1(x,y,\frac{1}{4})$, $R_4(1\frac{1}{2}-x,y+\frac{1}{2},\frac{1}{4})$ and those of the second orbit: $R_2(1-x,1-y,\frac{3}{4})$, $R_3(x-\frac{1}{2},\frac{1}{2}-y,\frac{3}{4})$, with x and y according to Table 4.4. The symbols (f,a) used to describe the basis functions for the R orbits correspondent to F and A for a two-atoms orbit.*

of the propagation vector can be obtained[89, 90]. The basic functions vector results of the symmetry analysis for the P*bnm* space group are listed in Tables 7.1.

If the antiferromagnetically stacked ferromagnetic layers are ordered in different modes[2] other reflections are produced as listed in Table 7.1[97].

---

[2]The different modes of the A-type can be understood of the different directions of the ferromagnetic sheets (stacked antiferromagnetically) in the crystal allowed by symmetry. For example in the A-mode the sheets are in the [001]-plane while in the F-mode the sheets lay in the [110]-plane

	h+k =	l =	first reflections
A-mode	2n	2n+1	(001)
G-mode	2n+1	2n+1	(101) / (011)
C-mode	2n+1	2n	(100) / (010)
F-mode	2n	2n	(110)*

Table 7.2: *Extinction rules for the possible magnetic peaks of an antiferromagentic Pbnm system with table showing the allowed reflections for the different magnetic modes of the A-type antiferromagneticorder. The asterisk at the F-mode marks that this rule is also valid for the nuclear magnetic peaks.*

7.2 Magnetic Structure $x = 0.45$ and 0.55

Polycrystalline measurements showed for the $x = 0.45$ sample a coexistence of an A-type antiferromagnetic component and a modulated spin order. On the other hand the samples $x \geq 0.5$ showed only the modulated spin order and no A-type antiferromagnetic component. As discussed in Section 7 the inversion symmetry breaking cycloidal spin order in TbMnO$_3$ is the origin for exhibiting a polarization orthogonal to the unit vector of the cycloid and the propagation direction as described in Formula (1.11). Having a modulated spin order now poses the question of the detailed magnetic structure and the possibility of having a spontaneous polarization by magnetic order.

To answer this question, single crystal sample neutron diffraction data from $x = 0.45$ and 0.55 were measured at D10 in four circle high flux mode. Although the crystals had more than one grain, the orientation of the grains allowed to collect data at high quality to determine the magnetic structure of these crystals. The nuclear data at 60 K was analyzed to resolve the structural data including scale factors and extinction parameters. Overall 351 independent nuclear Bragg reflections consisting of 502 individual reflections were collected at 60 K, 30 K and 2 K for determination of the crystallographic parameters such as the exact atomic positions and ratios and the absorption and extinction parameters. At temperatures of 30 K and 2 K additionally magnetic peaks were measured with a wavenumber of $q = 0.275$ for $x = 0.55$ and $q = 0.20$ for $x = 0.45$. They consisted of 196 independent magnetic Bragg reflections consisting of 459 individual reflections. Analytical absorption corrections for each reflection were made while the data collection by the measurement program, whereas the raw data was analyzed with Racer and the analysis of the integrated magnetic and nuclear intensities was performed with the FullProf suite.

The samples were cooled and a q-scans at 2 K were performed. At this temperature the most intense magnetic reflections were satellites of the A-type (001) reflection split along the b-direction. The sample $x = 0.55$ did not show any A-type (001) reflection whereas the sample $x = 0.45$ still showed these magnetic reflections. Further details on the magnetic ground state and temperature dependent structure and the development of those reflections with x and temperature are discussed in Section 6. The wavenumbers were determined resulting in the value for $x = 0.55$ of $q = 0.275$ while for $x = 0.45$ the wavenumber was $q = 0.20$.

At 2 K the most intense magnetic reflections with the determined wavenumbers showed an A mode order with a considerably weaker G mode component. No F mode reflections could be detected while a small C mode order component did neither improve nor disturb the analysis for the $x = 0.45$ sample. As the components must be orthogonal and a G_x component is already existent the C_x component though is excluded. A satisfactory fit for the data could be performed by using two independent ireps, namely $\Gamma_2 \otimes \Gamma_3 = (C_x, F_y, A_z) \otimes (G_x, A_y, F_z)$, resulting in a agreement index $R = 16.6$ and weighted $R_w = 30.1$ for $x = 0.45$ and in $R = 6.5$ and weighted $R_w = 13.2$ for $x = 0.55$. Figure 7.1 shows the observed and the calculated magnetic structure factors (F^2_{obs} and F^2_{calc}) at 2 K.

The higher R values for $x = 0.45$ are due to an overlap of the satellites and the remaining according A-type reflection at this measurement on many reflections. All values of the magnetic components are displayed in the Table 7.3. This means that at 2 K the spins are located in the bc-plane with an small component along the a-direction, just above the detection threshold. In both samples the ratio of $\frac{A_z}{A_y}$ is $\frac{2}{3}$. The moments of the Nd^{3+} ions only showed moments in Γ_2 at a_x and f_y with the same incommensurate ordering as known from DyMnO$_3$[98].. The phase between the two ireps was found to be best fitting the data at a phase shift at $\frac{\pi}{2}$ resulting in a cycloidal spin density wave in bc-plane propagating in the b-direction. The Figure 7.2 visualizes the different spin order in the two samples at the two temperatures. As $P_s \propto m_y m_z (e_x \times k)$, with k as the wavevector, this result confirms the observation of the polarization along $P||c$ at 2 K. The moments of the Nd^{3+} ions only showed moments in Γ_2 at a_x and f_y.

Measuring the reflections at 30 K only the A mode in the b-direction is left as well as the neodymium moments. This means that at 30 K only a spin density wave along the b-direction is left and a phase transition must happen between 2 K and 30 K. The spin order of Nd^{3+} did not change much in th $x = 0.45$ sample but nearly vanished at the $x = 0.55$ sample.

The development of the magnetic structure of this model though can be described that with cooling the paramagnetic phase of this crystal is by a great extend replaced by an A_y mode of an A-type antiferromagnetic phase forming a collinear spin density wave in b-direction. At lower temperatures a second A_z mode appears with a phase shift of $\frac{\pi}{2}$ producing an elliptical cycloid (with the approximately 1.5 times higher value along the b-direction) contained within the ab plane and the Mn spins rotating around the c axis. The values of the Mn spins in the commensurate cycloidal structure along the b and c axis for the $z = 0$ and $z = \frac{1}{2}$ layers of the $x = 0.45$ and $x = 0.55$ samples are given in Figure 7.3. This is equivalent to the results for TbMnO$_3$, that also exposes an elliptical cycloid with unit vector and propagation vector pointing in the same direction, and can explain the

7.2. Magnetic Structure $x = 0.45$ and 0.55

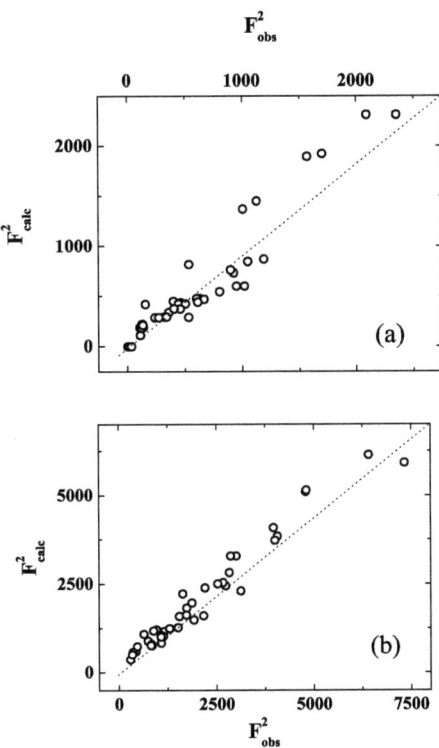

Figure 7.1: The comparison between the observed and the calculated magnetic structure factors (F^2) at 2 K is shown here for the analysis of the Mn spins reflections of the samples $x = 0.45$ (a) and $x = 0.55$ (b).

electric polarization along the c axis (as $P \propto a \times b$) in those crystals at these temperatures as it will be described in Section 5. The other active G_x mode with values between 0.38 and 0.47 have no effect on the cycloidal order but is responsible for a small canting.

These measurements completely confirm the results so far. Pyroelectric measurements proved that these samples are ferroelectric below 24 K with a polarization along the c-direction. Temperature dependent measurements showed different patterns for the samples $x = 0.45$ and 0.55. The sample $x = 0.55$ showed a stable wave number with temperature with a kink in the temperature dependent intensity at T_s. The measurements in this

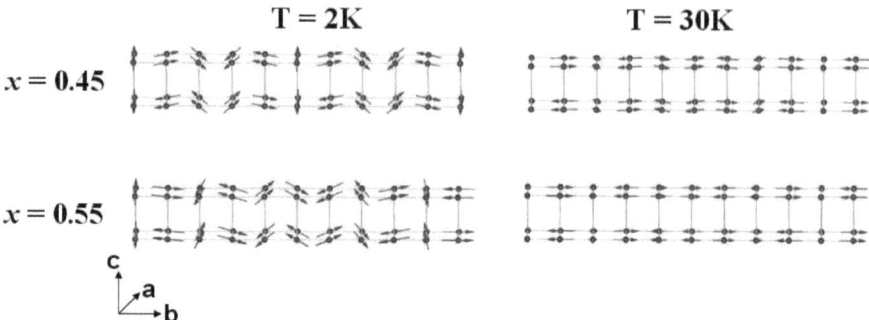

Figure 7.2: The modulation of the manganese spins of the samples $x = 0.45$ and 0.55 at temperatures of 2 K and 30 K.

	2K		30 K	
	$x = 0.45$	$x = 0.55$	$x = 0.45$	$x = 0.55$
Mn^{3+}				
A_y	2.8 (1)	3.34 (5)	2.7 (1)	2.84 (5)
A_z	1.8 (2)	2.1 (1)	0.5 (2)	0.0 (0)
G_x	0.38 (7)	0.47 (4)	0.0 (0)	0.0 (0)
Nd^{3+}				
a_x	1.8 (4)	1.2 (3)	1.7 (3)	0.6 (4)
f_y	0.8 (4)	1.0 (2)	0.9 (3)	0.6 (2)
R	16.6	6.5	9.1	6.2
R_w	30.1	13.2	16.6	9.1

Table 7.3: Moments and orientation of the Mn^{3+} and Nd^{3+} spins at 2 K and 30 K. All values are in $\frac{\mu_B}{mol}$. R is the agreement index, R_w is the weighted agreement index.

chapter prove that below T_N a collinear incommensurate magnetic component along the easy axis appears that cannot induce a ferroelectric polarization as $P \propto m_y \cdot m_z$. But with the onset of the second orthogonal component, that shows a phase shift of $\frac{\pi}{2}$, below T_s a cycloidal structure is developed what is consistent to the known multiferroic $RMnO_3$.

The discussion of the analysis of the sample $x = 0.45$ is more difficult. The analysis of the magnetic order in Section 6 showed that the sample has three magnetic components: an A-type antiferromagnetic component and two incommensurate components, one with a longer and one with a shorter wave vector and the longer one with a dependence on the A-type. For the analysis of the cycloidal order the integrated intensities of both incommen-

7.2. Magnetic Structure x = 0.45 and 0.55

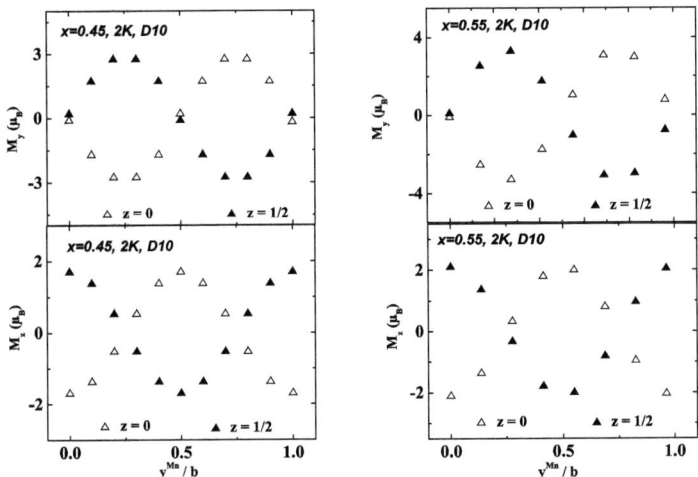

Figure 7.3: Results of the magnetic structure refinements of the Mn spins in $x = 0.45$ (left column) and 0.55 (right column) samples at 2 K showing the variation of the M_y (top row) and M_z (bottom row) components for the ions located at $z = 0$ (empty triangles) and $z = \frac{1}{2}$ (filled triangles) along the magnetic propagation vector with the components along a and b being out of phase.

surate components had to be done as both peaks highly overlapped. Further the diffuse reflections of the A-type structure also had a small overlap with the incommensurate peaks. This may count as a reason that the analysis of the reflections in this sample is not as good as in the $x = 0.55$ sample, resulting for example in reflections with high integrated intensity values ($F^2_{obs} > 1000$ in Figure 7.1 (a)) that could not be perfectly optimized, increasing the value of the agreement index. However the analysis of the magnetic structure using the magnetic reflections matches perfectly the results of the measurements of the last chapters.

Summary and Conclusions

Chapter 8
Summary and Conclusions

This thesis shows that it is possible to synthesize $Nd_{1-x}Y_xMnO_3$ orthorhombic single phase polycrystalline sample in the region of $0 \leq x \leq 0.6$ and single crystals in the region of $0.3 \leq x \leq 0.55$. These materials mimic to a great extend the complex electromagnetic properties of the rare-earth manganite $RMnO_3$ series from $R = $ Nd to Dy as well as the solid solution $Eu_{1-x}Y_xMnO_3$, despite the fact that the measured polarization are smaller here than in the other compounds. Different neutron diffraction measurements as a function of composition from the collinear A-type to the cycloidal ferroelectric compositions show a region of co-existence in x for both types of order parameters. This observation would suggest that this cross over is of first order with some added degrees of complexity, as other magnetic wavevector are observed in T,x space in this region.

8.1 The $Nd_{1-x}Y_xMnO_3$ Magnetic Phase Diagram

Neutron diffraction measurements on polycrystalline samples show that in the system of $Nd_{1-x}Y_xMnO_3$ the structural tuning of the nearest and next nearest magnetic interactions can be held responsible for the changes of T_N and the impact on the magnetic order below T_N. This is also known for the $Eu_{1-x}Y_xMnO_3$ series. Neutron powder diffraction on the polycrystalline samples show the co-existence of two magnetic components. Those two magnetic reflections although close in 2Θ are resolvable. However the information that can be determined from the powder data is limited when dealing with complex magnetic order. Despite this limitation there are several conclusions that can be drawn from these data:

- With increasing x (decreasing $\langle r_A \rangle$), T_N for the A-type order decreases in a similar fashion as in the $RMnO_3$ series.

- With increasing x the A-type order is replaced by an incommensurate spin order, where in both cases the spins are aligned parallel to the $b-$axis.

- There is a co-existence region between $0.4 < x < 0.45$ that exhibits both the collinear A-type phase and the incommensurate phase. This region separates the A-type and incommensurate phase fields.

- The ratio of intensities of the incommensurate and commensurate A-type magnetic reflections varies with x,

- An induced ferromagnetic component of Nd-spins was found whose amplitude decreases with increasing Y content.

The growth of single crystal samples enabled a more detailed analysis of the magnetic structures of this solid solution as well as the measurements of physical properties. For $x=0.3$ to 0.35 the A-type antiferromagnetic order was confirmed as previously found for NdMnO$_3$ and the powder measurements. Here the Mn-spins are aligned along the b-axis forming ferromagnetic layers that are antiferromagnetically coupled.

Turning the attention to the incommensurate magnetic order found for the higher x compositions, similar characteristics to the A-type antiferromagnetic order were found. Here the incommensurate propagation vector is also along the b-axis and the Mn-spins have a b-axis component and the antiferromagnetic stacking along the c-axis is maintained. The single crystals measurements for the $x = 0.45$ and 0.55 compositions the present of two magnetic transitions on cooling, the first at T$_N$ to a spin density wave with spins pointing along the b-axis and a second transition at T$_s$ to a cycloidal spin order. The cycloidal phase forms by the development of an additional c-axis component that is $\frac{\pi}{2}$ out of phase with the b-axis component. In the case of a cycloidal structure in the bc plane and a propagation along the b-direction a spontaneous polarization along the c-direction is predicted according to Formula (1.11).

The variation of magnetic order in the solid solution Nd$_{1-x}$Y$_x$MnO$_3$ tracks closely that of the RMnO$_3$ series. In both TbMnO$_3$ and DyMnO$_3$ compounds a bc cycloid is found and indeed the values of $\langle r_A \rangle$ for these compounds are comparable to those of the $x = 0.55$ sample. Compounds at the phase boundary between A-type antiferromagnetic and incommensurate order such as EuMnO$_3$ and GdMnO$_3$ have shown some evidence of an incipient A-type antiferromagnetic order with an easy axis along the b-direction[99]. This A-type antiferromagnetic order is possibly suppressed with the application of a magnetic field, as ferroelectricity is only found under magnetic field in this compound[100]. In the Nd$_{1-x}$Y$_x$MnO$_3$ solid solution indeed was found that the magnetic order between the A-type antiferromagnetic and incommensurate order co-exist over a certain region in composition ($0.4 < x < 0.45$) for slightly lower values of $\langle r_A \rangle$ than those of $R = $ Gd. The behavior that was found in those two compounds in terms of magnetic phase co-existence therefore may be indicative of the behavior of the $R = $ Gd and Eu compounds.

8.2 Multiferroic Properties

Performing temperature dependent measurements of the dielectric constant ϵ along the principle crystallographic axes reveal a clear transition along ϵ_c at about 24 K for crystals

8.2. Multiferroic Properties

	$x = 0.45$	$x = 0.55$
$m_y (\mu_B)$	2.8	3.34
$m_z (\mu_B)$	1.8	2.1
$m = \|m_y \cdot m_z\| (\mu_B)$	5.0	7.0
$P_c \left(\frac{\mu C}{m^2}\right)$	83	119
$R = \frac{m}{P_c}$	16.6	17.0

Table 8.1: *Ratio R of the product of the magnetic cycloid at 2 K to the strength of the polarization.*

of $x \geq 0.45$ what is comparable to the multiferroic RMnO$_3$. When applying high magnetic fields along the principle axes the transition in ϵ_c can be suppressed while a ϵ_a appears, indicating the formation of a P_c that can flop to P_a under field. This happens for $H\|b$ and $H\|c$ in contrast to TbMnO$_3$ and DyMnO$_3$, where $H\|a$ and $H\|b$ are necessary, while for R = Tb $H\|c$ suppresses the magnetic cycloid and therefore ferroelectricity and in R = Dy there is no effect.

Magnetization measurements are in agreement with the neutron data and the transitions measured in the dielectric data. For example in the higher x compounds an antiferromagnetic transition is found at T$_N$ with the easy axis along b direction indicative of the collinear spin density wave, while at 24 K another magnetic transition is indicated in magnetization measurements along the c-direction, indicative of the additional m_z component of the spin cycloid. Ferroelectric polarization measurements are in agreement with these data. A $P\|c$ appears for $x = 0.45$ whose magnitude at low temperature increases with x. This spontaneous polarization can be flopped to $P\|a$ with either $H\|a$ and $H\|b$ as also indicated by the measurements of the dielectric constant. The magnitude of the polarization is proportional to the product of the magnetic moments of the cycloidal spins as $P_s \propto m_y \cdot m_z$. If the ratio R as $\frac{m_y \cdot m_z}{P\|c}$ was refined, it was found that it is essentially the same for both the $x = 0.45$ sample that exhibits phase co-existence and the $x = 0.55$ sample that does not (Table 8.1). This indicates that the landau coupling constants are unchanged over this composition range and that the magnitude of the ferroelectricity is simply tuned by the size of the magnetic moments in the cycloid.

When applying a high magnetic field in the multiferroic Nd$_{1-x}$Y$_x$MnO$_3$ compounds, a flop of the polarization is induced. The driving field for TbMnO$_3$ and DyMnO$_3$ is along $H\|a$ and $H\|b$ resulting in $P\|c \rightarrow P\|a$, for Eu$_{0.6}$Y$_{0.4}$MnO$_3$ only $H\|a$ is known[101] resulting in a $P\|a \rightarrow P\|c$ while the multiferroic Nd$_{1-x}$Y$_x$MnO$_3$ is unique as the flop can be induced by a $H\|c$ but not $H\|a$, as displayed in Table 8.2.

In this work it was found that for the multiferroic Nd$_{1-x}$Y$_x$MnO$_3$ compounds the bc-cycloid is stable as it is also found for the R = Tb and Dy. This is contract to the multiferroic compounds of the Eu$_{1-x}$Y$_x$MnO$_3$ solid solution where the ab-cycloid is stable and give a zero magnetic field polarization along the c-axis. The stability of one type of cycloid

TbMnO$_3$, DyMnO$_3$	P_c	$\xrightarrow{H\|a, H\|b}$	P_a
Eu$_{0.6}$Y$_{0.4}$MnO$_3$	P_a	$\xrightarrow{H\|a}$	P_c
Nd$_{1-x}$Y$_x$MnO$_3$ ($x \geq 0.45$)	P_c	$\xrightarrow{H\|b, H\|c}$	P_a

Table 8.2: *Flopping direction of the polarization with applied magnetic field on different multiferroic compounds.*

over the other has been discussed by Mostovoy [49] in terms of the anisotropy provided by the rare earth spins and for Eu$_{1-x}$Y$_x$MnO$_3$ by Mochizuki and Furukawa[61]. It is argued that the natural state of the cycloid with out external perturbations is in the ab-plane but the anisotropy of the R-spins rotates it to the bc-plane. This appears to be consistent with the general behavior found in these manganites as in the compound mentioned above only an ab-cycloid is found only for the non magnetic Eu compounds. Therefore in the Nd$_{1-x}$Y$_x$MnO$_3$ solid solution the magnetic Nd ions provide enough anisotropy to also place the cycloid in the bc-plane as in the other magnetic R manganites. Indeed it is found that in the cycloidal phase of the x=0.45 and 0.55 compounds there is a induced Nd order in the ab-place even when the A-site magnetic sublattice is significantly diluted. This would indicate that the Mn-Nd interaction is significantly strong and possibly can provide sufficient anisotropy to stabilize the bc-cycloid.

The impact of a significant Nd-Mn interaction may also be reflected in the differences in the behavior of the cycloid in magnetic field. In the R = Tb and Dy manganites a $H\|a$ or $H\|b$ field result in the flopping of the cycloid from the bc to the ab place and thus also flopping the direction of the polarization. For the Nd$_{1-x}$Y$_x$MnO$_3$ series a different behavior is found where a $H\|a$ has little effect and a $H\|b$ and $H\|c$ fields result in the flop of the polarization. Again it is suggested that this change in the magnetic field behavior arises from the changes in the magnetic anisotropy introduced by the Nd ion compared to those of Dy and Tb.

8.3 Phase co-existance

One of the new results of this work is the observation of a phase co-existence between the A-type antiferromagnetic and incommensurate phase at the region that separate the purely A-type antiferromagnetic and cycloidal phase. This type of phase co-existence has been suggested for the GdMnO$_3$ [102] and Eu$_{0.8}$Y$_{0.2}$MnO$_3$[103] however never directly probed due to the high neutron absorption cross section of Gd. This suggest that one possible aspect that control such phase co-existence is the competing nearest and next nearest neighbor interactions that are tuned with $\langle r_A \rangle$. An other aspect is the quenched disorder that is introduced into the lattice when the larger size ion Nd is replaced with the smaller ion Y.

Summarizing the results of these measurements it is found that for the $x = 0.3$ compound only the A-type antiferromagnetic order, while for $x = 0.35$ to 0.45 the co-existence of A-type antiferromagnetic order with incommensurate order and for $x = 0.5$ and 0.55 only incommensurate order exists. The coherence length of these there magnetic orders varies with x, becoming smaller for the A-type antiferromagnetic order with increasing x and larger for the incommensurate order. While there appear complex transitions within the incommensurate phases between one stable wavevector to the other, the general trend suggest that the tuning of the magnetic interactions in $Nd_{1-x}Y_xMnO_3$ by varying $\langle r_A \rangle$ and in the presence of a quenched disorder results in a first order transition from a collinear A-type antiferromagnetic phase to a incommensurate spin cycloid. This observation is in contrast to the Monte Carlo simulations of the stability of the A-type antiferromagnetic and incommensurate order where no phase co-existence is predicted [48]

Looking more closely to the crystal chemical aspects, the difference of $Nd_{1-x}Y_xMnO_3$ not only to $RMnO_3$ but also to $Eu_{1-x}Y_xMnO_3$ is the is the significant size mismatch of the two A-site atoms. This may introduce a random potential into the perovksite lattice that in turn can modulate magnetic interactions significantly on the local scale. Such random potential or quenched disorder was discussed in phase separation models of the CMR or charge doped manganites and it is generally regarded to lead to phase separation at the boundary between the two competing phases[104] here the A-type and spin density wave magnetic phases. The effect of small compositional fluctuations that have a large effective spacial range due to the relaxation of the lattice around them, can favor one magnetic phase over the other, so the statistical distribution of the different A-site ions may influence the magneto-electric properties leading to different effects. Although no structural phase separation is observed it is reasonable to assume that in areas where only one kind of A-site atom arrange statistically in small clusters they may influence the magnetic properties to some degree[105].

8.4 Conclusion

In this thesis it was shown that:

- The there is a $Nd_{1-x}Y_xMnO_3$ solid solution where the orthorhombic perovskite phase is stable up to $x = 0.55$.

- For lower doped compounds $x \leqslant 0.3$ the A-type antiferromagnetic magnetic order is stable while Nd spins are induced to order ferromagnetically.

- For the composition region $0.35 \leqslant x < 0.5$ there is a phase co-existence between A-type antiferromagnetic and incommensurate order. This phase co-existence is partly driven by the size mismatch of the Nd and Y ions in the solid solution and may amplify the competition between nearest and next nearest Mn magnetic interactions. Within this setting the transition from A-type antiferromagnetic order to an incommensurate cycloid as a function of x is suggested to be first order.

- For the $x = 0.45$ and 0.55 compound it is confirmed that the incommensurate order arises from a bc Mn spin cycloid at 2 K with (likely) induced Nd order in the ab-plane.

- For composition $0.45 \leqslant x \leqslant 0.55$ it is found that the development of ferroelectricity that is associated with the onset of a Mn cycloidal magnetic order. The ferroelectricity that is generated by this order is consistent with the antisymmetric DM interaction model. The incommensurate magnetic wavevector that is ascribed to the cycloidal phase is of longer wavelength that the incommensurate phase found in the lower doped region of the co-existence region.

- A magnetic field applied along the b- and c-axis in the ferroelectric compositions of this solid solution results in the flop of the polarization from the c- to the a-axis, presumably driven by the flop of the cycloidal plane from the bc to the ab-plane.

- The variation of magnetic and ferroelectric properties, that are summarized in Figure ??, follow the general trend mapped out for the RMnO$_3$ series of perovskite manganites.

8.4. Conclusion

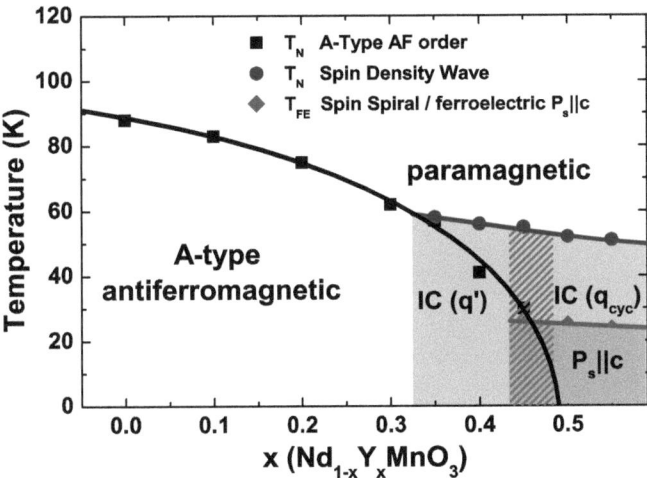

Figure 8.1: *Magnetic phase diagram of the series of $Nd_{1-x}Y_xMnO_3$ compounds from $x = 0.0$ to 0.55. Shown is T_N for the A-type antiferromagnetic component, T_N for the modulated spin components and the transition temperature T_{FE} of the cycloidal spin spiral causing a spontaneous polarization along the c-direction at 0 T. The existence of the modulated spin component is shown for longer wavelength (q') $q \approx 0.20$ and shorter wavelength (q_{cyc}) $q \approx 0.25$, respectively.*

Summary and Conclusions

Appendix A
Appendix A: Instruments Used

A.1 SQUID

The Superconducting Quantum Interference Device (SQUID) is used to measure very small magnetic fields using superconducting magnets. Attainable sensitivities of flux densities (10^{-12} T), of electrical current (10^{-12} A) and of electrical resistance (10^{-12} Ω) reflect the high accuracy of a SQUID. In superconductors the supercurrent is carried by pairs of electrons, known as Cooper pairs. Each pair can be treated as a single particle with a mass and charge twice that of a single electron, whose velocity is that of the center of mass of the pair. In a normal conductor the coherence length of the conduction electron wave is quite short due to scattering. Cooper pairs, however, are not scattered hence their wave functions are coherent over very long distances. This electron-pair wave retains its phase coherence over long distances and it is this characteristic which leads to interference and diffraction phenomena. As they are macroscopic manifestations of quantum interactions the phenomena are collectively termed 'Quantum Interference'. The phase of the electron-pairs can be affected not only by the current density but also quite strongly by an applied magnetic field. If two superconducting regions are kept totally isolated from each other the phases of the electron-pairs in the two regions will be unrelated. If the two regions are brought together then as they come close electron-pairs will be able to tunnel across the gap and the two electron-pair waves will become coupled. As the separation decreases the strength of the coupling increases. The tunneling of the electron-pairs across the gap carries with it a superconducting current as predicted by B.D. Josephson and is called JOSEPHSON-tunneling with the junction between the two superconductors called JOSEPHSON-junction.

The SQUID uses the properties of electron-pair wave coherence and Josephson Junctions to detect very small magnetic fields. The central element of a SQUID is a ring of superconducting material with one or more weak links as shown in Figure A.1. This produces a very low current density making the momentum of the electron-pairs small. The wavelength of the electron-pairs is thus very long leading to little difference in phase between any parts of the ring. If a magnetic field H is applied perpendicular to the plane of the ring, a phase difference is produced in the Cooper-pair wave. A small current is also induced to flow around the ring (J), producing a phase difference across the weak links

(1, 2). The phase difference due to the circulating current can either add to or subtract from that produced by the applied magnetic field. The circulating current has a periodic dependence on the magnitude of the applied field, a very small amount of magnetic flux. Detecting this circulating current enables the use of a SQUID as a magnetometer.

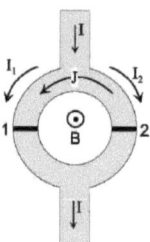

Figure A.1: *Schematic diagram of the superconducting ring in a SQUID magnetometer*

A.2 PPMS

As the determination of the magnetic properties is a central part of this work it was important to use an instrument that could the generate an stable external magnetic field with very low sample temperatures and allows to measure several physical properties of the sample under these conditions. This multifunctional instrument was a PPMS (**p**hysical **p**roperty **m**easurement **s**ystem), Model 6000 manufactured by Quantum Design as shown in Figure A.2. It allows temperatures from 1.9 K to 400 K and magnetic fields up to 14 T. This instrument was used for measurements of magnetic (VSM) and dielectric properties as well as for polarization measurements of the single crystal samples.

Figure A.2: *Photo of the Quantum Design PPMS.*

A.3 VSM

The VSM is one possible equipment of the PPMS. VSM stands for a **vibrating sample magnetometer** and measures the magnetic moment of a sample with the help of induced voltage. A change of the magnetic flux, according to Faraday's law,

$$\Phi = \int B dA \tag{A.1}$$

results in an induced voltage

$$U_{ind} = -\frac{d\Phi}{dt} \tag{A.2}$$

that can be measured in an VSM. The voltage U_{ind} is detected with so called pick-up coils and the change of Φ is achieved by vibrating the sample trough those coils. The sample and the pick-up coils are within an external field which is used to measure the field dependence of M. With z being the position of the sample within the pick-up coils the Equation A.2 can be written

$$U_{ind} = -\frac{d\Phi}{dz}\frac{dz}{dt} = CM sin(2\pi f t) \tag{A.3}$$

In Equation A.3 the assumption is made that the motion of the sample is sinusoidal with vibrational frequency f and C is a coupling constant. So by detecting the sinusoidal voltage induced by the pick-up coils the magnetization can be measured. The detectable moment reads 10^{-9} Am^2 to 0.04 Am^2 with an measurement accuracy better than $6 \cdot 10^{-9}$ Am^2 or 2%.

A.4 Dielectric Measurement Device

To measure the dielectric constant at different temperatures and external magnetic fields a tailor-made gadget was built for the PPMS. Here the sample, ideally very thin (1 mm) with a large area (100 mm^2) sputtered on both sides with a thin layer of gold, was clamped between two dies of sapphire having conducting surfaces connected to an external capacitance bridge (Andeen Hagerling, AH 2700A) as shown on Figure A.3.

The device allows to treat the plate with an alternating voltage to measure the capacitance for calculating the dielectric constant. Placing this device in the PPMS instrument allows to measure this constant with changing field and temperature.

Figure A.3: *Image of the sample holder for dielectric and polarization measurements in the PPMS instrument with a sample between the sapphire dies in $P\|b$ position.*

A.5 Polarization Measurement Device

The setup for measuring the polarization the cut, polished and sputtered crystals were clamped in the device mentioned Section A.4 and placed in the sample chamber of the PPMS. The crystal was cooled to a temperature above the ferroelectric transition (in most cases 25 K), than a voltage of 150 V was applied (with a Stanford PS350 High Voltage DC Power Supplies) and the crystal was cooled to 2.5 K. When reaching the temperature the voltage was replaced by an electrometer (Keithley 6517 System Electrometer) sensitive enough to measure a current of 0.01 pA. The crystal then was heated up to 35 K and the current was measured. The integration of the curve over time calculates to the charge on the crystal surface at 2.5 K that makes with the known crystal area the polarisation.

$$P_s = \frac{Q}{A} = \frac{\int_{t1}^{t2} I(t)dt}{A} \qquad (A.4)$$

A.6 E9

E9 is a neutron powder diffraction instrument at Helmholtz-Zentrum Berlin für Materialien und Energie T5 beam tube, a thermal flux about 1014 $\frac{n}{cm^2 s}$, a Ge monochromator and a wavelength of $\lambda_1 = 1.79714$ Å. The scattered beam is collected with a He$_3$ detector bank, a scetch of the instrument is shown in Figure A.4.

A.7 E4

E4 is also an instrument on the neutron facility at Helmholtz-Zentrum Berlin für Materialien und Energie used to measure single crystals on a two-axis arrangement under conditions of low temperatures and high magnetic fields. A vertically bent focusing monochromator creates an incident neutron beam with a wavelength of $\lambda = 2.44$ Å. The scattered

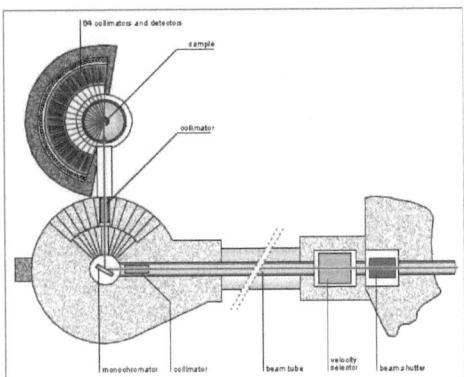

Figure A.4: *Schematic display of instrument E9 at the Helmholtz-Zentrum Berlin für Materialien und Energie. Thermal neutron beam coming from the reactor on the right is deflected at a monochromator and focused at the sample. The scattered neutrons are collected at a detector bank.*

neutrons are detected with a 2D detector, the collected data was analyzed with the LAMP[1] runtime environment v.6.4, a sketch of the instrument is shown in Figure A.5.

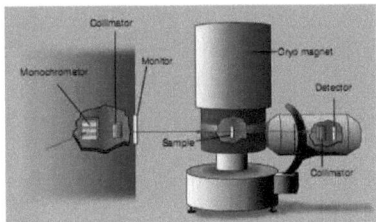

Figure A.5: *Sketch of the E4 instrument*

A.8 E5

E5 is a conventional four-circle diffractometer at the Helmholtz-Zentrum Berlin für Materialien und Energie. It is located at the beam tube R3 with a wavelength of 2.4 Å, a sketch of the instrument is shown in Figure A.6.

---

[1]LAMP stands for **L**arge **A**rray **M**anipulation **P**rogram, developed initially for the treatment of data obtained from neutron scattering experiments at the ILL

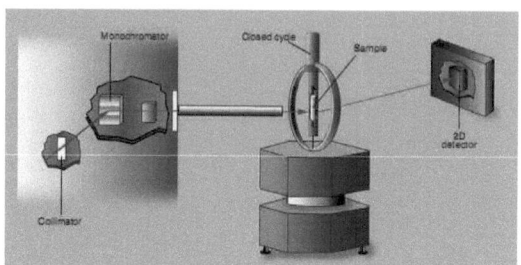

Figure A.6: Sketch of the E5 instrument.

A.9 V2

V2 is a triple-axis spectrometer installed at the cold neutron guide NL 1B at the Helmholtz-Zentrum Berlin für Materialien und Energie. The use of small incident neutron energies inherently gives good resolution properties. Here the wavelength of $\lambda = 2.47$ Å and a collimation of 40' was used. A sketch of the instrument is shown in Figure A.7.

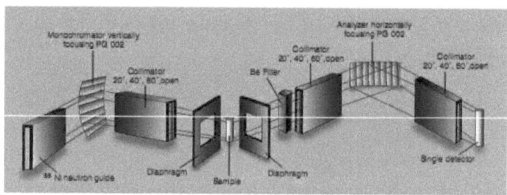

Figure A.7: Sketch of the V2 instrument.

A.10 Bruker

X-ray measurement: Bruker D8 Advance Reflection Mode X-Ray Diffractometer1, Cu (K_α) Beam, $\lambda_1 = 1.5406$ Å, $\lambda_2 = 1.54439$ Å(0.5%), 6 mm slit, 40V, 40A, $2\theta = 10 - 70°$, step size $0.05°$, 1.5 seconds per step, Bragg-Brentano geometry as shown in Figure A.8.

A.11 Laue

The Laue diffractometer produces a white x-ray beam in an x-ray tube focusing a spot on a single crystal and the diffraction of the scattering is detected by an image plate in

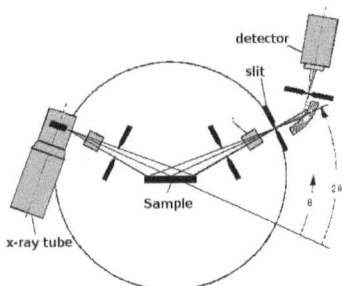

Figure A.8: At Bragg-Brentano geometry the sample is placed horizontally while the beam tube and the detector move in opposite directions on a circle.

back-scattering mode. The x-ray tube was a molybdenum tube operating at $U = 25$ kV and $I = 25$ mA. The image plate was read by a Fujifilm BAS-2500 scanner and the images were indexed[106] with OrientExpress v.3.4[2].

A.12 D10

The instrument D10 is located at the ILL (Institut Laue-Langevin) in Grenoble, France. It is a four-circle diffractometer with a unique four-circle cryostat for temperatures as low as 0.1 K, and offers high reciprocal-space resolution and low intrinsic background, to medium real space resolution. Figure A.9 shows the set-up of the instrument.

A.13 D20

The instrument D20 is also located at the ILL (Institut Laue-Langevin) in Grenoble, France. It is a high-intensity two-axis powder diffractometer with variable resolution. The complete diffraction pattern covers a scattering range of 153.6ř and can be obtained in seconds. The variable monochromator take-off angle, up to 120ř, opens the possibility to use this instrument in an high resolution mode with lower flux or a high flux mode at medium resolution. Figure A.10 shows the set-up of the instrument.

---

[2]Software for free use offered by the ILL, Grenoble at http://www.ccp14.ac.uk/ccp/web-mirrors/lmgp-laugier-bochu/

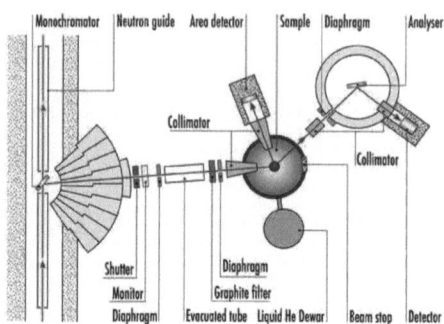

Figure A.9: *Schematic display of D10.*

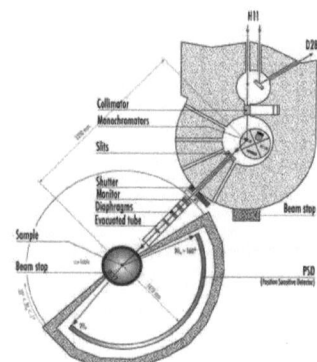

Figure A.10: *Schematic display of D20.*

A.14 Image Furnace

Single crystals were grown in a two mirror image furnace of NEC, Model SCI-MDH with water cooling and two 500 W lamps. In Figure A.11 the two hemispheres are visible to the left and the right as well as the feed rod holder from the top and the seed ro holder from the bottom. The digital camera for having a glance at the conditions at the molten zone is connected at the right hemisphere.

A.14. Image Furnace

Figure A.11: *The two mirror image furnace of NEC with hemispheres and rod holders visible.*

Summary and Conclusions

Appendix B
Appendix B: Detailled Figures

B.1 Results of Physical Properties Measurements

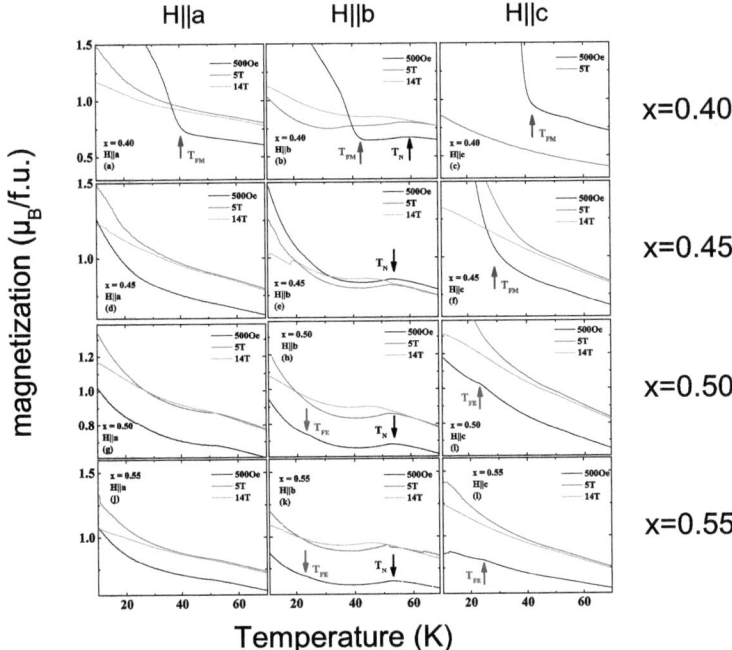

Figure B.1: *Temperature profiles of magnetization along the a, b, and c axis at magnetic fields of H = 500 Oe, 5 T and 14 T.*

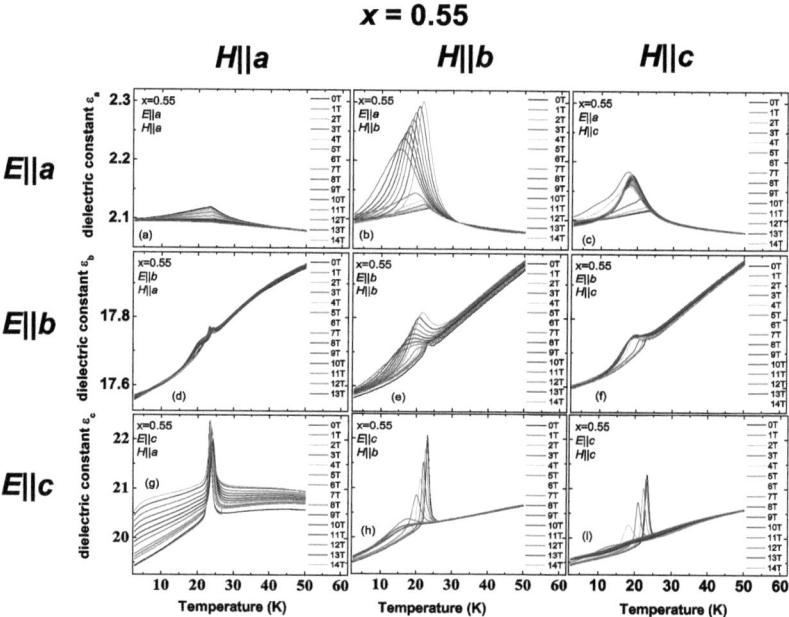

Figure B.2: Sample $x = 0.55$ temperature profiles of dielectric constant along the a, b, and c axis at magnetic fields of $H = 0$ T to 14T.

B.1. Results of Physical Properties Measurements

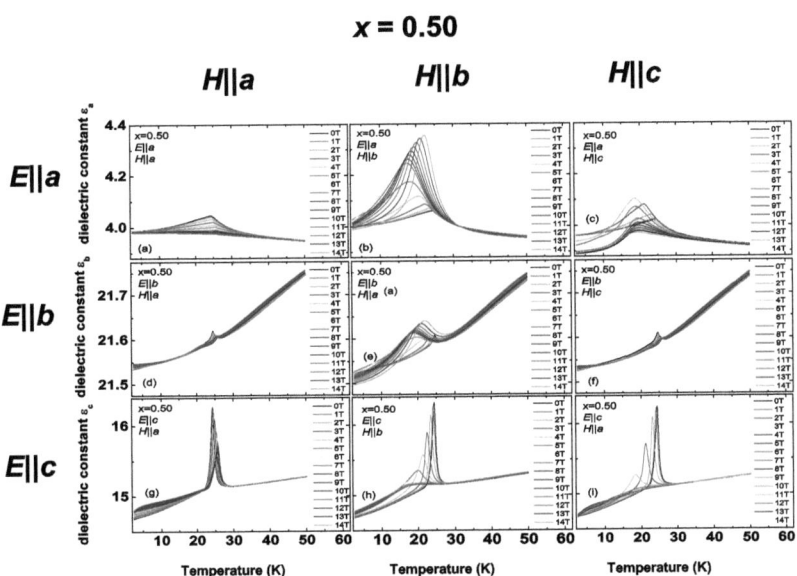

Figure B.3: Sample $x = 0.50$ temperature profiles of dielectric constant along the a, b, and c axis at magnetic fields of $H = 0$ T to 14T.

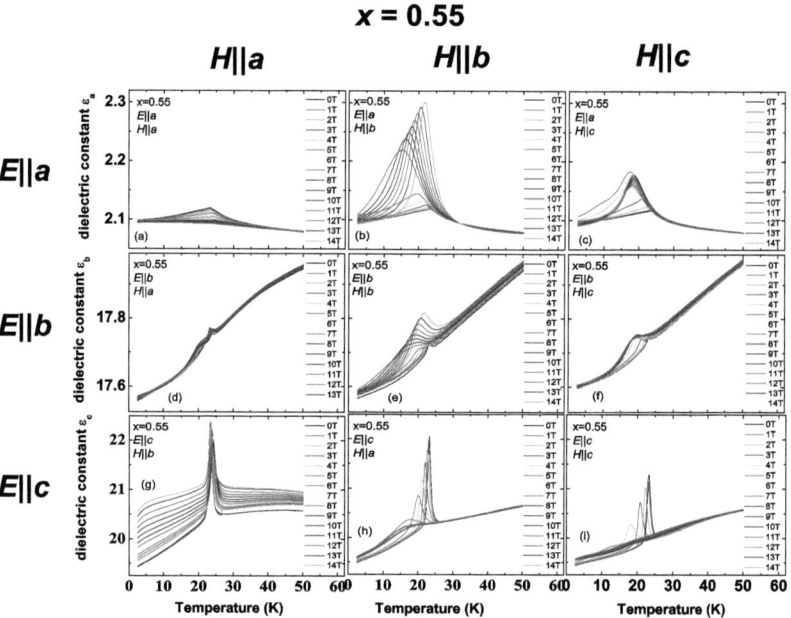

Figure B.4: Sample $x = 0.45$ temperature profiles of dielectric constant along the a, b, and c axis at magnetic fields of $H = 0T$ to $14T$.

B.1. Results of Physical Properties Measurements

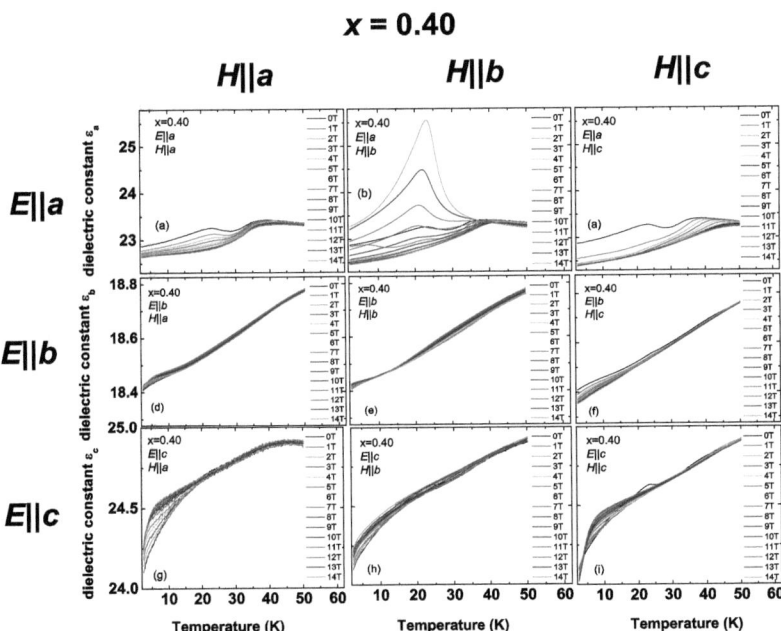

Figure B.5: Sample $x = 0.40$ temperature profiles of dielectric constant along the a, b, and c axis at magnetic fields of $H = 0$ T to 14T.

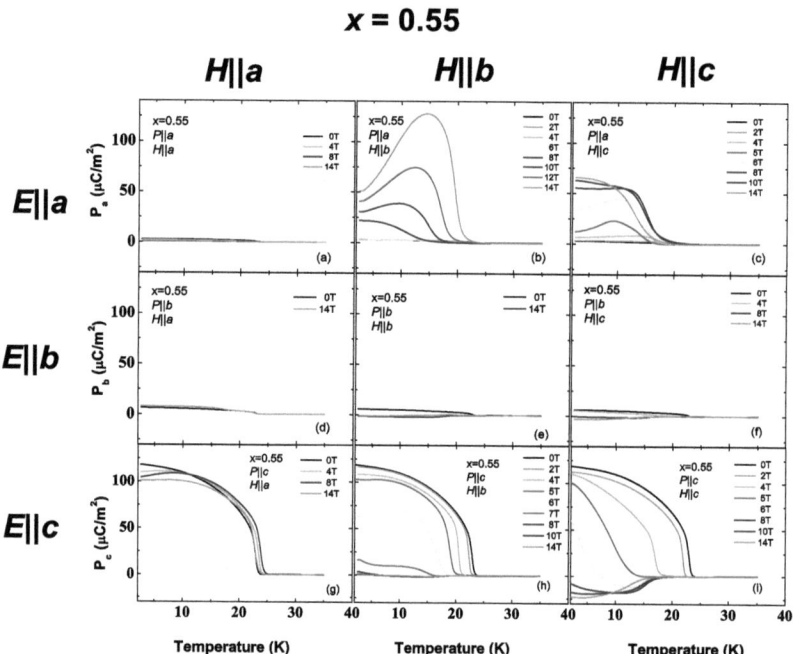

Figure B.6: Sample $x = 0.55$ temperature profiles of electric polarization along the a, b, and c axis at magnetic fields of $H = 0T$ to $14T$.

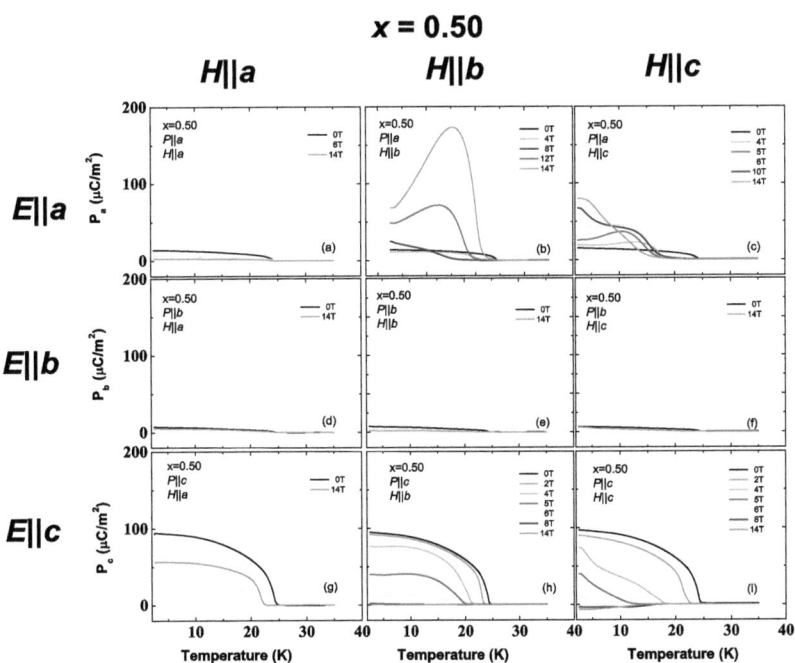

Figure B.7: Sample $x = 0.50$ temperature profiles of electric polarization along the a, b, and c axis at magnetic fields of $H = 0$ T to 14T.

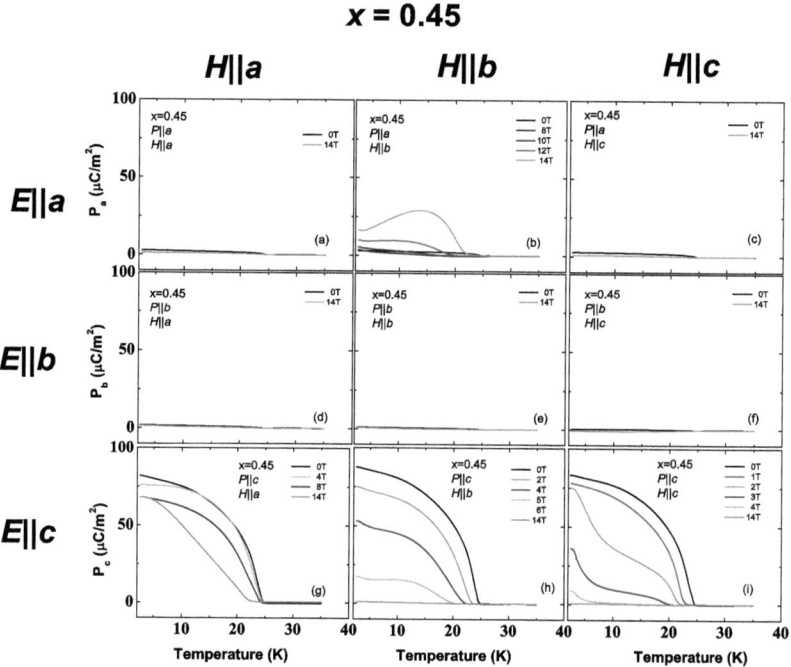

Figure B.8: Sample $x = 0.45$ temperature profiles of electric polarization along the a, b, and c axis at magnetic fields of $H = 0T$ to $14T$.

B.1. Results of Physical Properties Measurements

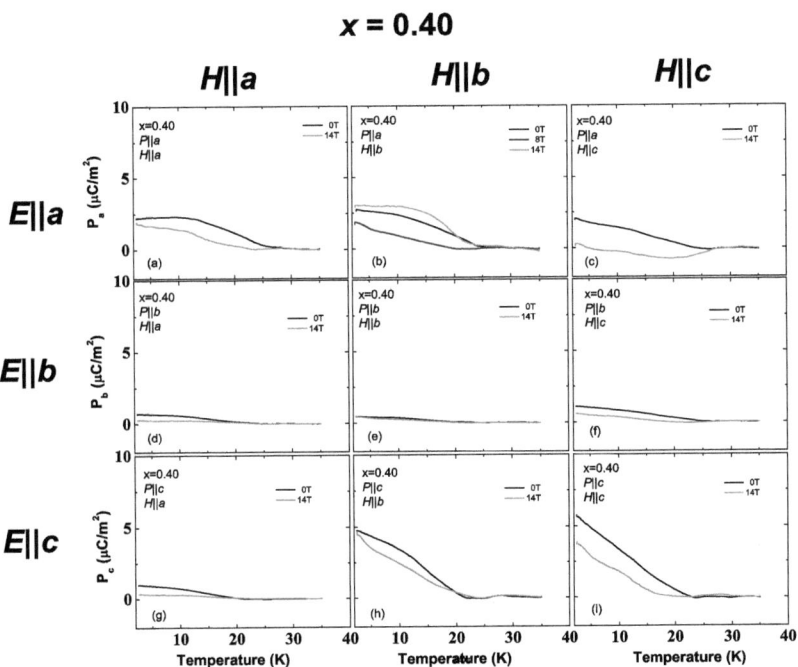

Figure B.9: Sample $x = 0.40$ temperature profiles of electric polarization along the a, b, and c axis at magnetic fields of $H = 0$ T to 14T. The error of the polarization is about $\pm 1.5 \frac{\mu C}{m^2}$.

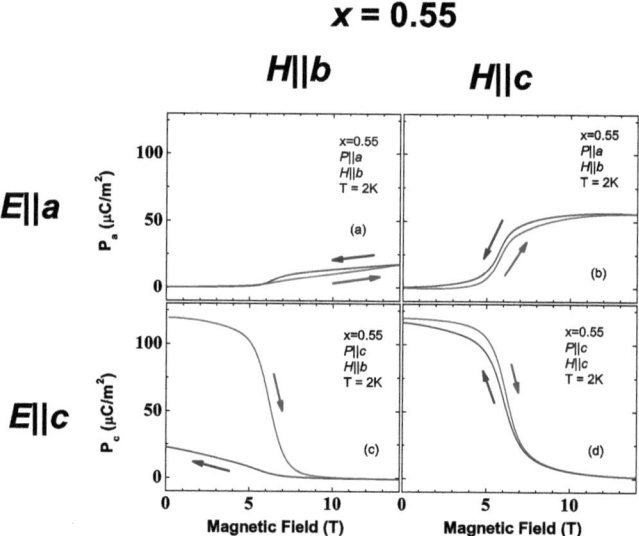

Figure B.10: Magnetic field profile of the sample $x = 0.55$ of polarization measured along the c and a axis with field sweep $H\|c$ and $H\|b$ from 0 T to 14 T and back to 0 T with 100 Oe/min at 2 K. Red lines are increasing field, blue lines are decreasing field.

B.1. Results of Physical Properties Measurements

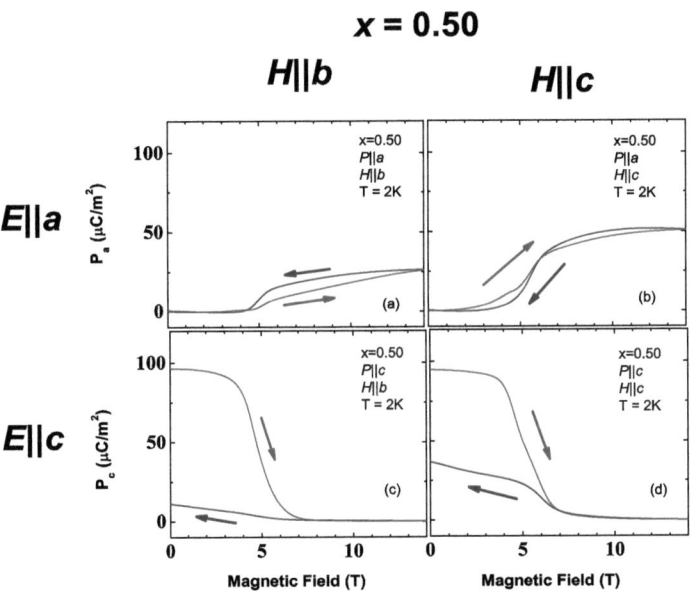

Figure B.11: Magnetic field profile of the sample $x = 0.50$ of polarization measured along the c and a axis with field sweep $H\|c$ and $H\|b$ from 0 T to 14 T and back to 0 T with 100 Oe/min at 2 K. Red lines are increasing field, blue lines are decreasing field.

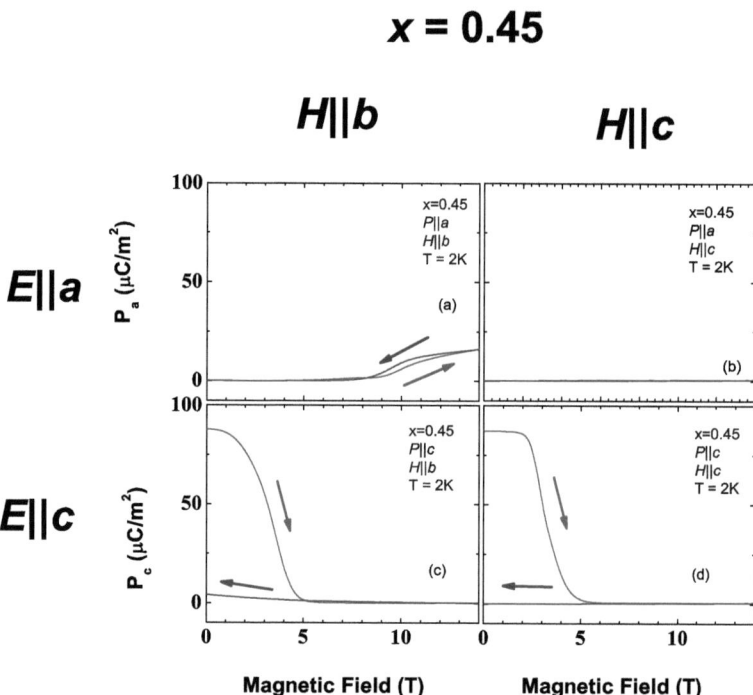

Figure B.12: *Magnetic field profile of the sample $x = 0.45$ of polarization measured along the c and a axis with field sweep $H\|c$ and $H\|b$ from 0 T to 14 T and back to 0 T with 100 Oe/min at 2 K. Red lines are increasing field, blue lines are decreasing field.*

Summary and Conclusions

Appendix C
Appendix C: Publications

S. Landsgesell, A. Maljuk, T. C. Hansen, O. Prokhnenko, N. Aliouane, and D. N. Argyriou, *Structural tuning of magnetism in the orthorhombic perovskite $Nd_{1-x}Y_xMnO_3$: Evidence for magnetic phase separation*, Physical Review B **80**, 014412 (2009)

H. N. Bordallo, E. V. Boldyreva, A. Buchsteiner, M. M. Koza, and **S. Landsgesell**, *Structure-Property Relationships in the Crystals of the Smallest Amino Acid: An Incoherent Inelastic Neutron Scattering Study of the Glycine Polymorphs*, Journal of Physical Chemistry **B 112**, 8748-59 (2008)

J. Strempfer, B. Bohnenbuck, I. Zegkinoglou, N. Aliouane, **S. Landsgesell**, M. V. Zimmermann, and D. N. Argyriou, *Magnetic-field-induced Transitions in multiferroic $TbMnO_3$ Probed by Resonant and Nonresonant X-Ray Diffraction*, Physical Review B **78**, 024429 (2008)

O. Prokhnenko, R. Feyerherm, E. Dudzik, **S. Landsgesell**, N. Aliouane, L. C. Chapon, and D. A. Argyriou, *Enhanced ferroelectric polarization by induced Dy spin order in multiferroic $DyMnO_3$*, Physical Review Letters **98**, 057206 (2007)

N. Aliouane, D. N. Argyriou, J. Strempfer, I. Zegkinoglou, **S. Landsgesell**, and M. V. Zimmermann, *Field-induced linear magnetoelastic coupling in multiferroic $TbMnO_3$*, Physical Review B **73**, 020102 (2006)

C. J. Milne, D. N. Argyriou, A. Chemseddine, N. Aliouane, J. Veira, **S. Landsgesell**, and D. Alber *Revised superconducting phase diagram of hole-doped $Na_x(H_3O)_zCoO_2 \cdot yH_2O$*, Physical Review Letters **93**, 247007 (2004)

R. Feyerherm, A. Loose, **S. Landsgesell**, and J. L. Manson, *Magnetic ordering in iron tricyanomethanide*, Inorganic Chemistry **43**, 6633-6639 (2004)

Bibliography

[1] Wadhawan, V. K. *Introduction to Ferroic Materials*. CRC Press, (2001).

[2] Levanyuk, A. P. and Sannikov, D. *Soviet Physics Uspekhi* **17**, 199–214 (1974).

[3] Newnham, R. E. and Cross, L. E. *Materials Research Bulletin* **9**(7), 927 (1974).

[4] Känzig, W. *Ferroelectrics and Antiferroelectrics in Solid State Physics: Advances in Research and Applications*, volume 4. Academic Press, (1957).

[5] Bokov, A. A. and Ye, Z.-G. *Journal of Materials Science* **41**, 31 (2006).

[6] Cross, L. E. *Ferroic Materials and Composites: Past, Present Ferroic Materials and Composites: Past, Present and Future*. Advanced Ceramics III. Elsevier Applied Science, (1990).

[7] Newnham, R. E. and Cross, L. E. *Materials Research Bulletin* **9**(7), 1021 (1974).

[8] Fiebig, M. *Journal of Physics D* **38**, R125–R152 (2005).

[9] Tokura, Y. *Science* **312**, 1481 (2006).

[10] Tokura, Y. *Journal of Magnetism and Magnetic Materials* **310**, 1145 (2007).

[11] Eerenstein, W., Mathur, N. D., and Scott, J. F. *Nature* **442**, 759 (2006).

[12] Smolenskii, G. A. and Chupis, I. E. *Uspekhi Fizicheskikh Nauk* **137**(3), 415–448 (1982).

[13] Jona, F. and Shirane, G. *Ferroelectric Crystals*. Dover, (1993).

[14] Schmid, H. *Ferroelectrics* **162**, 317–338 (1994).

[15] Hill, N. *Journal of Physics and Chemistry B* **104**, 6694–6709 (2000).

[16] Lines, M. E. and Glass, A. M. *Principles And Applications Of Ferroelectrics And Related Materials*. Oxford Univ. Press, (2001).

[17] Khomskii, D. *Bulletin of the American Physial Society* **C**, 21.002 (2001).

[18] Kimura, T., Goto, T., Shintani, H., Ishizaka, K., Arima, T., and Tokura, Y. *Nature* **426**(6962), 55–58 (2003).

[19] Kimura, T., Kawamoto, S., Yamada, I., Azuma, M., Takano, M., and Tokura, Y. *Physical Review B* **67**, 180401(R) (2003).

[20] Astrov, D. *Soviet Physics JETP-USSR* **11**(3), 708–709 (1960).

[21] St-Gregoire, P., Almairac, R., and Gesland, J. *Ferroelectrics* **67**(15-21), 15 (1986).

[22] Lottermoser, T., Lonkai, T., Amann, U., Hohlwein, U., Ihringer, J., and Fiebig, M. *Nature* **430**, 541–544 (2004).

[23] Zhang, Y., Yang, H. X., Guo, Y. Q., Ma, C., Tian, H. F., Luo, J. L., and Li, J. Q. *Physical Review B* **76**, 184105 (2007).

[24] Lonkai, T., Tomuta, D. G., Amann, U., Ihringer, J., Hendrikx, R. A., Tö bbens, D. M., and Mydosh, J. A. *Physical Review B* **69**, 134108 (2004).

[25] Lueken, H. *Angewandte Chemie* **120**, 8690 – 8693 (2008).

[26] Khomskii, D. *Journal of Magnetism and Magnetic Materials* **306**, 1 (2006).

[27] Cheong, S. and Mostovoy, M. *Nature Materials* **6**(1), 13–20 (2007).

[28] Goto, T., Kimura, T., Lawes, G., Ramirez, A., and Tokura, Y. *Physical Review Letters* **92**(25), 257201 (2004).

[29] Kimura, T., Lawes, G., Goto, T., Tokura, Y., and Ramirez, A. *Physical Review B* **71**, 224425 (2005).

[30] Hur, N., Park, S., Sharma, P., Ahn, J., Guha, S., and Cheong, S.-W. *Nature* **429**(6990), 392–395 (2004).

[31] Yamasaki, Y., Miyasaka, S., Kaneko, Y., He, J., Arima, T., and Tokura, Y. *Physical Review Letters* **96**, 249902 (2006).

[32] Goldschmidt, V. M. *Chemische Berichte* **60**, 1263 (1927).

[33] Norby, P. and Hanson, J. C. *Journal of Solid State Chemistry* **119**, 191 (1995).

[34] Goodenough, J. B. *Journal of the Physics and Chemistry of Solids* **6**, 287 (1958).

[35] Goodenough, J. *Physical Review* **100**, 564 (1955).

[36] Kanamori, J. *Journal of the Physics and Chemistry of Solids* **10**, 87 (1959).

[37] Mitchell, J., Argyriou, D., Potter, C., Hinks, D., Jorgensen, J., and Bader, S. *Physical Review B* **54**, 6172–6183 (1996).

[38] Mizokawa, T., Khomskii, D., and Sawatzky, G. *Physical Review B* **60**, 7309 (1999).

[39] Solovyev, I., Hamada, N., and Terakura, K. *Physial Review Letters* **76**, 4825 – 4828 (1996).

[40] Kimura, T., Ishihara, S., Shintani, H., Arima, T., Takahashi, K., Ishizaka, K., and Tokura, Y. *Physical Review B* **68**(6), 060403 (2003).

[41] Muñoz, A., Casáis, M. T., Alonso, J. A., Martínez-Lope, M. J., Martínez, J. L., and Fernández-Díaz, M. T. *Inorg. Chem.* **40**, 1020 (2001).

[42] Huang, Y., Fjellvåg, H., Karppinen, M., Hauback, B. C., Yamauchi, H., and Goodenough, J. B. *Chemistry of Materials* **19** (**8**), 2139 (2007).

[43] Okamoto, H., Imamura, N., Hauback, B. C., Karppinen, A., Yamauchi, H., and Fjevag, H. *Solid State Communications* **14**, 152–156 (2008).

[44] Pomjakushin, V. Y., Kenzelmann, M., Doenni, A., Harris, A. B., Nakajima, T., Mitsuda, S., Tachibana, M., Keller, L., Mesot, J., Kitazawa, H., and Takayama-Muromachi, E. *New Journal of Physics* **11**, 043019 (2009).

[45] Sergienko, I. and Dagotto, E. *Physical Review B* **73**, 094434 (2006).

[46] Chapon, L. C., Blake, G. R., Gutmann, M. J., Park, S., Hur, N., Radaelli, P. G., and Cheong, S.-W. *Physical Review Letters* **93**(17), 177402 (2004).

[47] Chapon, L. C., Radaelli, P. G., Blake, G. R., Park, S., and Cheong, S.-W. *Physial Review Letters* **96**, 097601 (2006).

[48] Dong, S., Yu, R., Yunoki, S., Liu, J.-M., and Dagotto, E. *Physical Review B* **78**, 155121 (2008).

[49] Mostovoy, M. *Physical Review Letters* **96**(6), 067601 (2006).

[50] Katsura, H., Nagaosa, N., and Balatsky, A. *Physical Review Letters* **95**(5), 057205 (2005).

[51] Bary'achtar, V. G., L'vov, V. A., and Jablonskii, D. A. *JETP Letters* **37**, 673 (1983).

[52] Stefanovskii, E. P. and Jablonskii, D. A. *Sovjet Journal of Low Temperature Physics* **12**, 478–480 (1986).

[53] Kenzelmann, M., Harris, A. B., Jonas, S., Broholm, C., Schefer, J., Kim, S. B., Zhang, C. L., Cheong, S.-W., Vajk, O. P., and Lynn, J. W. *Physical Review Letters* **95**(8), 087206 (2005).

[54] Dzyaloshinsky, I. *Journal of Physical and Chemical Solids* **4**, 241 (1958).

[55] Moriya, T. *Physial Review Letters* **4**, 228 (1960).

[56] Arima, T., Tokunaga, A., Goto, T., Kimura, H., Noda, Y., and Tokura, Y. *Physical Review Letters* **96**(9), 097202 (2006).

[57] Lawes, G., Harris, A. B., Kimura, T., Rogado, N., Cava, R. J., Aharony, A., Entin-Wohlman, O., Yildirim, T., Kenzelmann, M., Broholm, C., and Ramirez, A. P. *Physial Review Letters* **95**, 087205 (2005).

[58] Harris, A. B., Yildirim, T., Aharony, A., and Entin-Wohlman, O. *Physical Review B* **73**, 184433 (2006).

[59] Sergienko, I., Şen, C., and Dagotto, E. *Physical Review Letters* **97**, 227204 (2006).

[60] Skumryev, V., Ott, F. and Coey, J., Anane, A., Renard, J.-P., Pinsard-Gaudart, L., and Revcolevschi, A. *European Physical Journal B* **11**, 401–406 (1999).

[61] Mochizuki, M. and Furukawa, N. *Journal of the Physical Society of Japan* **78**, 053704 (2009).

[62] Xiang, H. J., Wei, S.-H., Whangbo, M.-H., and Da Silva, J. L. F. *Physical Review Letters* **101**, 037209 (2008).

[63] Malashevich, A. and Vanderbilt, D. *Physial Review Letters* **101**, 037210 (2008).

[64] Ishiwata, S., Kaneko, Y., Tokunaga, Y., Y., T., Arima, T., and Tokura, Y. (2009).

[65] Yamasaki, Y., Miyasaka, S., Goto, T., Sagayama, H., Arima, T., and Tokura, Y. *Physical Review B* **76**, 184418 (2007).

[66] Noda, K., Akaki, M., Kikuchi, T., Akahoshi, D., and Kuwahara, H. *50th Annual Conference on Magnetism and Magnetic Materials* **99**, 08S905 (2006).

[67] Hemberger, J., Schrettle, F., Pimenov, A., Lunkenheimer, P., Ivanov, V., Mukhin, A., Balbashov, A., and Loidl, A. *Physical Review B* **75**, 035118 Jan (2007).

[68] Goto, T., Yamasaki, Y., Watanabe, H., Kimura, T., and Tokura, Y. *Physical Review Letters* **72**, 220403 (2005).

[69] Brinks, H., Fjellvag, H., and Kjekshus, A. *Journal of Solid State Chemistry* **129**, 334 (1997).

[70] Rietveld, H. *Acta Crystallographica A* **22**, 151 (1967).

[71] Rietveld, H. *Journal of Applied Crystallography* **2**, 65 (1969).

[72] Ok, K., Chi, E., and Halasyamani, P. *Chemical Society Reviews* **35**, 710–717 (2006).

[73] Kamegashira, N. and Miyazaki, Y. *Materials Research Bulletin* **19**, 1201–1206 (1984).

[74] Cherepanov, V., Barkhatova, L., Petrov, A., and Voronin, V. *Journal of Solid State Chemistry* **118**, 53 (1995).

[75] Atsumi, T. and Kamegashira, N. *Journal of Alloys and Compounds* **252**, 67 (1997).

[76] Atsumi, T., Ohgushi, T., and Kamegashira, N. *Journal of Alloys and Compounds* **238**, 35 (1996).

[77] Hashimoto, T., Ishizawa, N., Mizutani, N., and Kato, M. *Journal of Crystal Growth* **84**, 207 (1987).

[78] Balbashov, A. M., Karabashev, S. G., Mukovskiy, Y. M., and Zverkov, S. A. *Journal of Crystal Growth* **167**, 365–368 (1996).

[79] Mori, T., Aoki, K., Kamegashira, N., Shishido, T., and Fukuda, T. *Materials Letters* **42**, 387 (2000).

[80] Mori, T., Kamegashira, N., Aoki, K., Shishido, T., and Fukuda, T. *Materials Letters* **54**, 238–243 (2002).

[81] Meng, J., Che, P., Mori, T., Sasai, R., Fujita, K., Satoh, H., Aoki, K., Kamegashira, N., Shishido, T., and Fukuda, T. In *Frontiers od Solid State Chemistry*, 97–101. World Scientific Publ. Co. PTE ltd., (2002).

[82] Landsgesell, S., Maljuk, A., Hansen, T. C., Prokhnenko, O., Aliouane, N., and Argyriou, D. N. *Physical Review B* **80**, 014412 (2009).

[83] Hemberger, J., Brando, M., Wehn, R., Ivanov, V., Mukhin, A., Balbashov, M., and Loidl, A. *Physical Review B* **69**, 064418 (2004).

[84] Wu, S., Kuo, C., Wang, H., Li, W., Lee, K., Lynn, J., and Liu, R. *Journal of Applied Physics* **87**, 5822 (2000).

[85] Mukhin, A., Ivanov, V., Travkin, V., and Balbashov, A. *Journal of Magnetism and Magnetic Materials* **1139**, 226–230 (2001).

[86] Geller, S. *The Journal of Chemical Physics* **24**, 1236–1239 (1956).

[87] Goodenough, J. *Physical Review* **100**, 564 (1955).

[88] Kajimoto, R., Yoshizawa, H., Shintani, H., Kimura, T., and Tokura, Y. *Physical Review B (Condensed Matter and Materials Physics)* **70**(1), 012401 (2004).

[89] Bertaut, E. *Journal of Magnetism and Magnetic Materials* **24**(3), 267 (1981).

[90] Brinks, H. W., Rodríguez-Carvajal, J., Fjellvåg, H., Kjekshus, A., and Hauback, B. C. *Physical Review B* **63**(9), 094411 (2001).

[91] Hukushima, K. and Nemoto, K. *Journal of the Physical Society of Japan* **65**, 1604–1608 (1996).

[92] Aliouane, N., Argyriou, D., Strempfer, J., Zegkinoglou, I., Landsgesell, S., and v. Zimmermann, M. *Physical Review B* **73**(2), 020102 (2006).

[93] Aliouane, N., Schmalzl, K., Senff, D., Maljuk, A., Prokes, K., Braden, M., and Argyriou, D. N. *Physical Review Letters* **102**, 207205 (2009).

[94] Pimenov, A., Rudolf, T., Mayr, F., Loidl, A., Mukhin, A. A., and Balbashov, A. M. *Physical Review B* **74**, 100403 (2006).

[95] Katsura, H., Balatsky, A. V., and Nagaosa, N. *Physial Review Letters* **98**, 027203 (2007).

[96] Strempfer, J., Bohnenbuck, B., Zegkino, I., Aliouane, N., Landsgesell, S., v. Zimmermann, M., and Argyriou, D. N. *Physical Review B* **78**, 024429 (2008).

[97] Quezel, S., Tcheou, F., Rossat-Mignod, J., Quezel, G., and Roudaut, E. *Physica B+C* **86-88**, 916–918 (1977).

[98] Prokhnenko, O., Feyerherm, R., Dudzik, E., Landsgesell, S., Aliouane, N., Chapon, L. C., and Argyriou, D. N. *Physical Review Letters* **98**(5), 057206-4 (2007).

[99] Ferreira, W. S., Moreira, J., Almeida, A., Araujo, J. P., Tavares, P. B., Mendonca, T. M., Carvalho, P. S., and Mendonca, S. *Ferroelectrics* **368**, 345–351 (2008).

[100] Ribeiro, J. L. *Ferroelectrics* **368**, 352–357 (2008).

[101] Murakawa, H., Onose, Y., Kagawa, F., Ishiwata, S., Kaneko, Y., and Tokura, Y. *Physical Review Letters* **101**(19), 197207 NOV 7 (2008).

[102] Kadomtseva, A., Y.F., P., Vorob'ev, G., Kamilov, K., Pyatakov, A., V.Y., I., Mukhin, A., and Balbashov, A. *JETP Letters* **81**, 19–23 (2005).

[103] Danjoh, S., Jung, J.-S., Nakamura, H., Wakabayashi, Y., and Kimura, T. *Physical Review B* **80**, 180408(R) (2009).

[104] Burgy, J., Mayr, M., Martin-Mayor, V., Moreo, A., and Dagotto, E. *Physical Review Letters* **87**, 277202 (2001).

[105] Friebel, C. *Acta Crystallographica A* **36**, 259–265 (1980).

[106] Laugier, J. and Filhol, A. *Journal of Applied Corystallography* **16**, 281–283 (1983).

I want morebooks!

Buy your books fast and straightforward online - at one of world's fastest growing online book stores! Environmentally sound due to Print-on-Demand technologies.

Buy your books online at
www.morebooks.shop

Kaufen Sie Ihre Bücher schnell und unkompliziert online – auf einer der am schnellsten wachsenden Buchhandelsplattformen weltweit! Dank Print-On-Demand umwelt- und ressourcenschonend produziert.

Bücher schneller online kaufen
www.morebooks.shop

KS OmniScriptum Publishing
Brivibas gatve 197
LV-1039 Riga, Latvia
Telefax: +371 686 204 55

info@omniscriptum.com
www.omniscriptum.com

Printed by Books on Demand GmbH, Norderstedt / Germany